Decoding the Mechanisms of Antikythera Astronomical Device

Jian-Liang Lin · Hong-Sen Yan

Decoding the Mechanisms of Antikythera Astronomical Device

Springer

Jian-Liang Lin
Department of Mechanical Engineering
National Cheng Kung University
Tainan
Taiwan

Hong-Sen Yan
Department of Mechanical Engineering
National Cheng Kung University
Tainan
Taiwan

ISBN 978-3-662-48445-6 ISBN 978-3-662-48447-0 (eBook)
DOI 10.1007/978-3-662-48447-0

Library of Congress Control Number: 2015950021

Springer Heidelberg New York Dordrecht London

Printed on acid-free paper

Springer-Verlag GmbH Berlin Heidelberg is part of Springer Science+Business Media (www.springer.com)

Preface

The Antikythera astronomical calculating device invented in the ancient Greece between 150–100 BC was discovered from the Antikythera shipwreck of the Mediterranean in 1900–1901. The device was the oldest astronomical calculator for the function of indicating the date of Egyptian calendar, displaying the motions of the Sun, the Moon, and the five planets, showing moon phases, calculating the calendars, and predicting eclipses. The interior structure of mechanism is composed of gear trains with six subsystems corresponding to its functions. Nevertheless, the device was absolutely strange to everyone in the era of the discovery. None of known ancient artifacts bears any resemblance to this excavated mechanical device, and the historical records or literatures about such a geared device were not available. Owing to the damages caused by excavation in some fragments, subsystems are incomplete, unclear, and even lost. Decoding the mechanisms and reconstructing the device are much more difficult, even with advanced modern image technology.

This book presents a unique systematic design methodology to decode the interior mechanisms of the Antikythera device. The historical background, the surviving evidence, and the existing reconstruction works of the device are introduced, and the historical development of astronomical achievements and various astronomical instruments is investigated. By utilizing the methodology based on the conceptual design of modern mechanisms, all feasible designs of the six lost/incomplete/unclear subsystems subject to the standards of science and technology in the time period are synthesized as illustrated examples, and 48 feasible designs of the complete interior mechanisms are presented. Such a design methodology provides not only a logical tool applying the knowledge of mechanical engineering for the reconstruction designs of the Antikythera device, but also an innovative research approach for identifying the original structure of mechanisms in the future.

The book is organized in such a way that it can be used for research, science education, and self-study. It can be used as a supplementary note for readers who are interested in the Antikythera device. Chapter 1 outlines the historical development of Western astronomy, including cosmology, calendars, astronomical events, and astronomical theory. Chapter 2 introduces the ancient astronomical instruments and analyzes the designs corresponding to some special functions. Chapter 3 introduces the historical background, known functions, and structure analysis of the Antikythera mechanism. Chapter 4 describes the existing reconstruction works. Chapter 5 presents the design methodology and compares the authors' reconstruction designs with other available ones. Chapters 6–10 are the reconstruction designs of the calendrical subsystem, the lunar subsystem, the solar subsystem, the planetary subsystem, and the moon phase display device, respectively. Chapter 11 explains the geometric constraints and detailed designs of assembly work and presents all feasible designs of complete interior structure along with a 3D model.

The authors would like to express their gratitude to those who had been of continuous assistance, especially the members of Creative Mechanism Design Research and Education Laboratory in the Department of Mechanical Engineering at National Cheng Kung University (Tainan, Taiwan), as well as the Ministry of Science and Technology (Taipei, Taiwan) for the financial support under Grant NSC 99-2221-E-006-253.

The authors believe that this book will meet the needs of academic research, archaeological research, and teaching in the reconstruction designs of ancient machinery and creative designs of modern mechanisms. Comments and suggestions for the improvement and revision of the book will be highly appreciated.

Tainan, Taiwan

Jian-Liang Lin
Hong-Sen Yan

Contents

Chapter 1
A Sketch of Ancient Western Astronomy

Long time ago, the cyclic motions of the Sun, the Moon, and several moving heavenly bodies were found. Many attentions were paid to these astronomical phenomena for making careful observations and keeping records. They were contributed to not only defining the time systems, such as day, month, and year, for actions and celebrations of human, but also explaining the connection between the human and the stars even to develop the astrology. Indeed, all sorts of accumulated knowledge nourished the embryo of astronomy. Ancient Greek astronomy, between 700 and 400 BC, was an outstanding branch of Western astronomy and contributed many brilliant achievements, including the forming of cosmology and the geometrical theories of celestial bodies. The purpose of this chapter is to study ancient Western astronomical developments, starting with the introduction of Western astronomy and ends with the astronomical achievements of the Hellenistic Age. In addition, several relevant civilizations, including the Egyptian civilization, the Mesopotamian civilization, and the Aegean civilization, calendars, special astronomical cycles, and astronomical theories to describe the motions of celestial bodies by geometrical models are introduced. The following content is the basis of investigation of the known astronomical devices. It is helpful for readers to understand Chaps. 3 and 4 for the known Antikythera astronomical device and its reconstruction work. Also, it is closely contributed to Chaps. 6–10 for synthesizing all possible reconstruction designs of the incomplete and lost mechanisms of this invention.

1.1 Historical Development of Western Astronomy

Astronomy that deals with the study of celestial objects and phenomena is one of the old academics has subjects and sprung up in several civilizations around the world. Although the definite original time of astronomy can hardly be proven, the

J.-L. Lin and H.-S. Yan, *Decoding the Mechanisms of Antikythera Astronomical Device*, DOI 10.1007/978-3-662-48447-0_1

well-known civilizations in the world, respectively, possessed their development of astronomy according to their excavations and historical remains. By investigating the available historical literatures and the archaeological discovery, it is believed that the Egyptian civilization, the Mesopotamian civilization, and the Aegean civilization should have played the crucial roles in the development of the later Greek astronomy. Furthermore, some great works among these civilizations could be inherited as forming schools or techniques. In what follows, the astronomical developments and achievements of these civilizations that effected the Greek astronomy are introduced. And, the forming process of the cosmology in Greek is also described.

1.1.1 Egyptian Civilization

The astronomy of the Egyptian civilization was built on the requirements of agriculture. In order to accurately control the timing for seeding and irrigation, the Egyptians needed a time system to successively measure time to tell them when to do what. A particular calendar similar to the modern calendar was built, and the Egyptian calendar is the earliest lunisolar calendar. Although the time length of a year adopted in this calendar is equal to 365.25 days, this calendar was established in accordance with the Sirius, and not the Sun. Certainly, the Egyptian astronomy benefited from their outstanding achievements and detailed records of astronomical observation. In addition to the building of the calendar, one important invention of the Egyptian astronomy was the definition of the zodiac, the orbit of the Sun, as shown in Fig. 1.1. The zodiac of the Egyptian astronomy was divided into 36 intervals for different stars with the period of 10 days. The relevant achievement reflected that the observation of constellations was highly developed. This concept of zodiac could be used to distinguish the seasons much precisely [1–10].

In ancient Egypt, there were many different drawings and glyphs to describe the universe. They visualized the structure of universe as a regular box, and the Earth and the Heaven were the bottom and the top of the box, respectively. The celestial *Nile* was the heavenly river crossing the universe, and the path of the celestial *Nile* was called the path of the *Milky Way* that was the path of souls. According to different historical records, the sky could be held up by 4 pillars or be supported by a pair of twin mountains, the eastern mountain called "*Manu*" and the western mountain called "*Bakhuf*." Whatever descriptions, these pillars of Heaven might be the same as the twin mountains. Furthermore, the Egyptian astronomy was a mythical astronomy. Figure 1.2 illustrates one of the various cosmologies of the Egyptian civilization [11]. The Egyptians saw the Earth as a flat expanse of land personified by the god—*Geb*, the sky as an arched and stretched body, like a bridge, of the goddess *Nut*. The goddess *Nut*'s hands lie in the western horizon, and her legs lie in the eastern horizon. They were separated by *Shu*, the god of air [1, 11–15].

Fig. 1.1 Zodiac of Egyptian civilization. Reprinted from Ref. [10], copyright 2015, with permission from Louvre Museum, Paris

Fig. 1.2 Universe of ancient Egyptian mythical astronomy [11]

Ancient Egyptians' different cosmologies derived their explanations for sunrise and sunset, or day and night. One belief indicated that the ancient Egyptians believed the Sun to rise from the mountain of *Bakhu* and to set in the mountain of

Manu. Another belief stated that the god of the Sun, *Ra*, traveled on his morning (solar) boat over the sky in the day and moved in the netherworld on his evening boat through the night. Still another belief stated that the goddess *Nut* swallowed the Sun at sunset and the Sun traveled through the body of *Nut* in the night. *Nut* expelled the Sun at sunrise and starts the day. In whatever cosmologies, these theories that explained sunrise and sunset were parallel [1, 6, 12–15]. Furthermore, for the changes of seasons, the ancient Egyptians stated that the Sun traveled close to or away from the Earth along the celestial river to generate four seasons.

In a word, the achievements of the Egyptian astronomy were remarkable, but they usually explained these results by spiritual viewpoints to match their belief and religion. Indeed, the Egyptian astronomy was a mythical astronomy that had its outstanding achievements on astronomical observation [6–15].

1.1.2 Mesopotamian Civilization

Mesopotamia indicates the land of two rivers, Tigris River and Euphrates River. In the modern geographical position, it is located largely in Iraq, northeastern Syria, southeastern Turkey, and southwestern Iran. It was believed to be the earliest region with the agricultural actions. This environment with fertile natural resources had gestated many ancient civilizations. Sumers, Babylonians, and Assyrians had built their empires successively in this region. In addition, the developed cultures were inherited one by one, especially for the observation records of the astronomical phenomena, which resulted in the highly developed astronomy of the Mesopotamian civilization.

The astronomical development of the Mesopotamian civilization was from the definition of time for the arrangement of festivities. A day, the basic unit of time, was defined as the period between two successive sunrises, and the beginning of day was at sunset. A month was defined as the changing period of the moon phase. A year was equal to 354 days or 12 months, including six 30-day long months and six 29-day short months. Furthermore, the leap month was presented to adjust the consistence between the seasons and the calendar. This arrangement of the calendar was regarded as one of the lunisolar calendars [3–6, 16].

Precise observations of the Mesopotamian civilization not only built a systematic and accurate calendar with its astronomical work, but also investigated the Heaven. Astronomical events could be calculated with considerable accuracy, and the solar and lunar years could be reconciled by intercalating months. The most important was to identify the motions of the Sun, the Moon, and the five planets including Venus, Mercury, Mars, Jupiter, and Saturn. At that time, the existence of the zodiac and distinguished between the planets and the fixed stars had already been formulated. The zodiac of the Mesopotamian civilization was divided into twelve intervals for various constellations, a concept of the zodiac that is still in use today. The five planets were observed and their synodic periods were also calculated exactly. Furthermore, the eclipses could be calculated owing to the

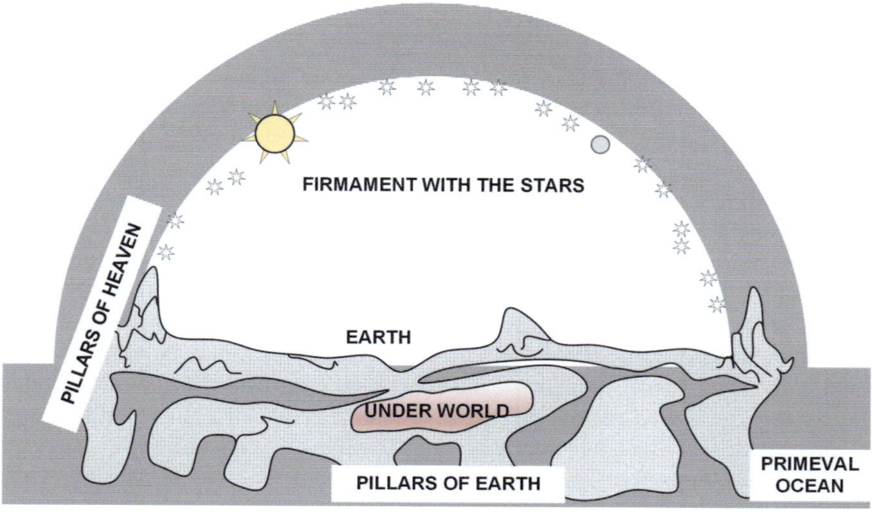

Fig. 1.3 The cosmology of Babylonian astronomy [22]

understanding of the nodes between the lunar and the solar orbits. The Saros cycle was a famous example to describe and predict the period of eclipses. Furthermore, the cosmology of the Mesopotamian civilization was built, as shown in Fig. 1.3, and included the Earth, the Heaven, and the ocean. The Earth was in the center surrounded by the ocean from up to down. Most of these outstanding scientific achievements of astronomy, even the mythical astronomy, were understood by decoding the excavated tablets of the civilization [8, 17–22].

Basically, the Babylonian astronomers devised the combinations of the arithmetical sequences to predict the motions of the observable celestial bodies. The achievements of the Babylonian astronomy are regarded as mathematic astronomy. In addition, it is proven that the astronomy of the Babylonian civilization eventually exercised a powerful influence on the development of Greek astronomy [23].

1.1.3 Aegean Civilization

Aegean civilization, which existed in the region bordering the Aegean Sea consisting of the Crete, the Cyclades, and some other islands, and the Greek mainland, was called the first civilization of Europe. Its time period, dating from the late Bronze Age of 1400 BC to the ancient Greece of 31 BC, consists of four major ages including the Age of Minoan and Mycenaean civilization, the Dark Age, the Classical Age, and the Hellenistic Age [1, 2, 24]. The following is the introduction of the astronomical achievements developed in each age.

1.1.3.1 Minoan and Mycenaean Civilization

No doubt, the title of "the first civilization of Europe" should be bestowed to the Minoan civilization of Crete, and the Mycenaean civilization of mainland Greece and the island and coast of the Aegean Sea [25]. According to the archaeological findings, it is confirmed that the Minoan civilization, which was earlier than the Mycenaean civilization, had indeed developed their astronomical investigations. The symbols for celestial objects were frequently depicted in the excavations of religious contexts. The most common evidence was a variety of solar and stellar symbols [25, 26].

One of the representative archaeological works of the Minoan civilization is the Minoan clay disk, as shown in Fig. 1.4. The excavated disk is presumed as a ritual calendrical object that shows the most important cycles of celestial bodies and probably was used in relation to the Minoan festival calendar. The disk, about 40 cm across, has five concentric circles consisting of small holes. The numbers of these holes, from the center of the disk rim, are 1, 15, 24, 38, and 61. Their functions are speculated as follows: the innermost one hole represented the day that should be skipped; the 15 holes represented half of the synodic month of 29.5 days; the 24 holes represented the lunar year; and the 38 and 61 holes represented together the 99 synodic months. If the above speculation is correct, the calendar applied in the Minoan civilization should be the same as the Egyptian calendar [25].

1.1.3.2 Dark Age

The Dark Age of the Aegean civilization (1100–750 BC) referred to the emptiness of historical records and not to the gloomy achievements. After the fall of the Mycenae by Dorian invasion, the great civilization was destroyed. Afterward, the historical records and writing, which included the astronomical knowledge at

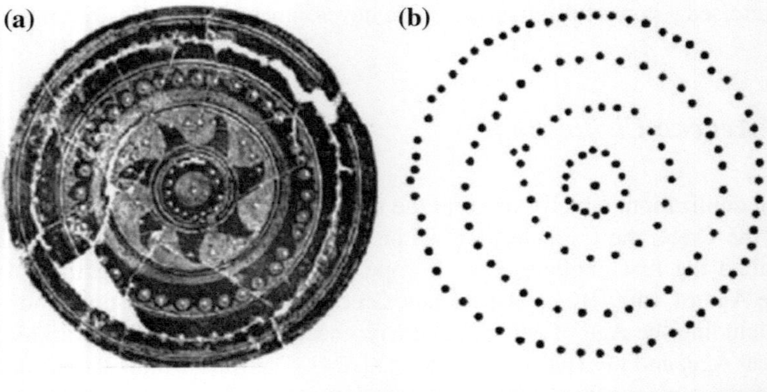

Fig. 1.4 Excavated Minoan clay disk, reprinted from Ref. [25], copyright 2015, with permission from Marianna Ridderstad. **a** Excavation. **b** Reconstructed concept

that time, were almost lost. Understanding the history of this troubled period was impeded by this unfortunate situation. Even so, it is believed that the achievements for technology, economy, and society should be existed and the festivities were still much frequent. This opinion could be proven by the historical record that the first game of Olympiad was held in 776 BC, and this celebration was still taken in the ancient Greek during the Hellenistic Age as a calendar [4, 25, 27].

1.1.3.3 Classical Age

The astronomical development of the Classical Age (750–323 BC) was the beginning of great achievements of ancient Greek astronomy. In this period, several schools with diverse principal concepts of the cosmology appeared and made their special contributions for people to understand what the universe was [1, 2, 6, 7, 28–35]. These astronomical attainments are really epic and some of which interacted with each other. In what follows, each school in this period is introduced.

1.1.3.4 Ionia School

Ionia school is referred to as the origin of astronomy in the Classical Age. Since the region of the ancient Greece was very close to Egypt and Babylonia, the astronomy of the Ionia school was influenced firstly by the Egyptian and the Mesopotamian civilizations. During this period, Thales of Miletus (624–547 BC) and Anaximander (610–545 BC) were two most outstanding astronomers of the school [1, 28].

Thales was the precursor of the Ionia school. His cosmology indicated that water was the principle of all things in the world, and the stars in the Heaven were fires. Since it was understood that the orbits of Moon and Sun did not coincide, he presented that the light of the Moon was reflected from the Sun [1, 2, 7]. However, in Anaximander's cosmology shown in Fig. 1.5, the structure of universe was a half ball rotating around the Polaris. The Earth was floating on and surrounded by water, air, and fire. Such a viewpoint overturned completely the early one of the Egyptian and the Mesopotamian civilizations that the Earth was infinite [1, 2, 6, 7, 28, 29].

1.1.3.5 Pythagoras School

The Ionia school was destroyed after the invasion of Persian. Then, the Pythagoras school was gradually developed. The most important representative of the school was Pythagoras of Samos (580–500 BC). He was the first Greek to explain the astronomy by the viewpoint of the geometrical mathematics. Based on the special phenomena, including the circular shape of shadow in the eclipse, the up and down of stars in the observer's movement, and the disappearing of ship from the sight of the observer, he proposed that the universe was a sphere [1, 2].

Fig. 1.5 Universe of Ionia school [35]

Certainly, Pythagoras' opinion also developed the cosmology of the Pythagoras school, as shown in Fig. 1.6. It was believed that the center of the spherical universe was a fire to provide the required light and heat. The Earth, the Sun, the Moon, and the planets rotated around the fire at a constant speed, and the (fixed) stars were stationary. But this viewpoint could not support the observable parallax of the fixed stars. In order to explain this problem logically, the cosmology was corrected further. The fire at the center of the spherical universe was replaced by the Earth. It should rotate around its own axis [1, 2, 7, 19, 28–30].

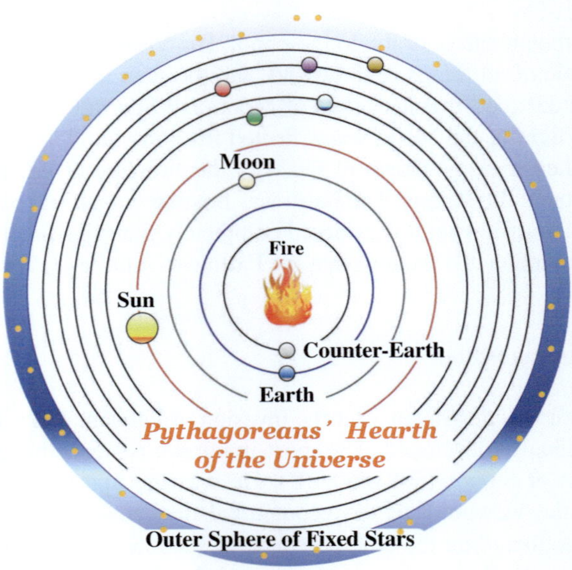

Fig. 1.6 Universe of Pythagoras school [30]

1.1.3.6 Plato School

The Plato school was developed at a prosperous time of the ancient Greece astronomy. It inherited the knowledge from the Ionia and the Pythagoras schools. Greek astronomer Plato (427–347 BC) was the founder of this school. His viewpoint indicated that the universe was a perfect sphere, and the anomaly motions of planets could be achieved through the combinations of several uniform circular motions. This viewpoint was demonstrated completely with the geometrical method by Eudoxus of Cnidus (409–356 BC).

Figure 1.7 is the structure of universe constructed by Aristotle. This concept of universe followed the Eudoxus' system of concentric spheres. Based on the description of theory, the center of the system was fixed as the Earth, and the outermost rotated one revolution per day as the fixed stars. Each of the Sun, the Moon, and the planets occupied its own sphere which was attached via its poles to a larger sphere that rotated about the other poles. For the motions of the Sun and the Moon, they could be described through the combinations of three concentric spheres, as shown in Fig. 1.8a. However, the motion of each planet should be demonstrated through the combinations of four concentric spheres, as shown in Fig. 1.8b. Therefore, in Eudoxus' system, 27 concentric spheres in all were used to describe the universe that Eudoxus had understood [28–34].

After Eudoxus, the geometrical method of the system of concentric spheres to describe the motions of celestial bodies was adopted generally. With the advanced precision of the observation and the increasing number of the known stars, the number of the spheres required in the systematic of concentric spheres should be on the rise. At least, it was confirmed that the number of the concentric spheres presented by Callippus was 56 [1, 28, 29].

Furthermore, Aristotle (384–322 BC), a famous Greek philosopher with great influence in the Western history, had several achievements in astronomy. His opinion indicated that the universe was finite and every heavenly body should be attracted by the Earth, the center of universe. Simultaneously, in addition to approving the theory of the concentric sphere, he further maintained that the

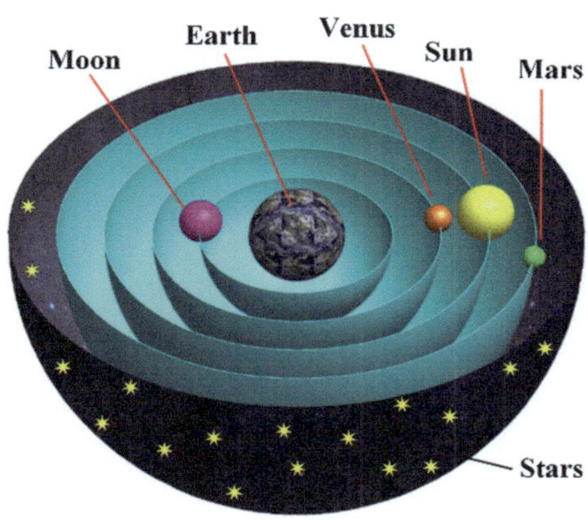

Fig. 1.7 Aristotle's structure of the universe [31]

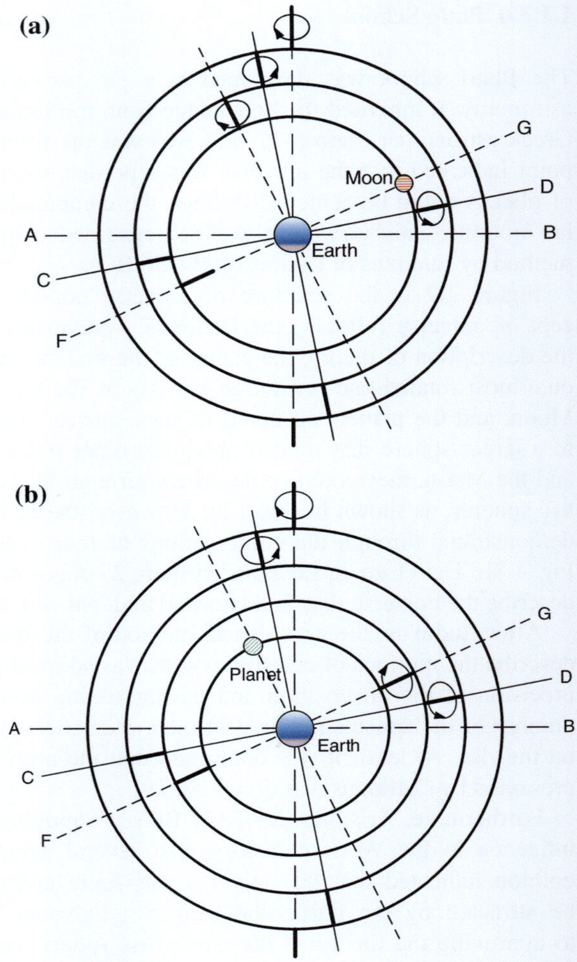

Fig. 1.8 Eudoxus' concentric spheres system for the motions of heavenly bodies [28]. **a** Motions of the Sun and the Moon. **b** Motions of the planets

outermost sphere for the fixed stars was driven by primum mobile. In accordance with the lengths of periods, the sequence of planets, distributing from the position nearest to the Earth, were the Moon, the Mercury, the Venus, the Sun, the Mars, the Jupiter, and the Saturn [28].

1.1.3.7 Hellenistic Age

The Hellenistic Age (323–31 BC) was the background of the Antikythera astronomical device. Several new geometrical models describing the astronomical motions were built successively. These sufficiently supported the design basis of the device. Furthermore, the study of Greek astronomy was advanced from the traditionally geometrical methods to the mathematical ones. The distances with the astronomical numbers could be measured.

Aristarchus of Samos (310–230 BC) was a prominent Greek philosopher and astronomer. His cosmology, which overturned completely the traditional viewpoint, indicated that the center of universe was the Sun. The Earth and the planets should rotate around the Sun with circular motions, and the Sun and the fixed stars were stationary. Then, the Earth took not only one revolution per year but also one rotation per day. Furthermore, he also maintained that the parallax of a fixed star was invisible, since the fixed star was farther from the Earth than the Sun. Except for the building of theory, Aristarchus was the first one attempting to measure the distances, respectively, among the Sun, the Moon, and the Earth through the geometrical relation of triangle, even though the calculating results were not exact. Undoubtedly, all of his astronomical works were quite creative at that period, yet were consistent with the cosmology known in modern time [28, 29, 32–34].

Some years after Eudoxus, the celestial system of the concentric spheres was overturned. Apollonius of Perga developed two equivalent geometrical systems, which were the epicyclic system and the eccentric system, respectively, to account for the motion of the Sun, as shown in Fig. 1.9a. They could be used for the demonstrations of the Moon and the planets. These two systems are introduced in Sect. 1.3. Simultaneously, he also realized crucially that the epicyclic system could account for the known retrograde motions of the planets if the epicycle rotated in the same sense as that with which its center rotated around the deferent, as shown in Fig. 1.9b. Once the radii and the speeds of the epicycle and the deferent that are introduced in Sect. 1.3 are chosen suitably, the motion, which is predominantly clockwise but with short-period rotating at opposite direction, is generated. Furthermore, he understood that the radii and the speeds of the epicycle and the deferent could determine the occurrences of stationary points for the planets [28].

Afterward, Hipparchus attempted to discuss Apollonius' geometrical models further. He took a mathematical attitude to handle the solar motion and calculated all parameters needed for Apollonius' eccentric theory, as shown in Fig. 1.10. Through his great contributions, the position of the Sun could be predicted with errors that could not be detectable by the naked eyes. As to the motion of the Moon, Hipparchus presented two equivalent lunar theories, which, respectively, used the epicyclic system and the eccentric system, to explain the lunar motion successfully. In his geometrical model of the Moon, not only the longitude but also the latitude is considered in detail. Then, through the study of the eclipse records at ancient time, he calculated precisely four special lunar periods, including the synodic month, the sidereal month, the draconitic month, and the anomalistic month. However, unlike the work of Apollonius, Hipparchus did not have any obvious achievement on the planetary motions. He might have realized that the planetary motions were too complex to be concluded into a satisfactory theory that accounts for them [28].

Hipparchus' astronomical achievements still included the definition of the degree of brightness of a star and the building of star table. Through the precise

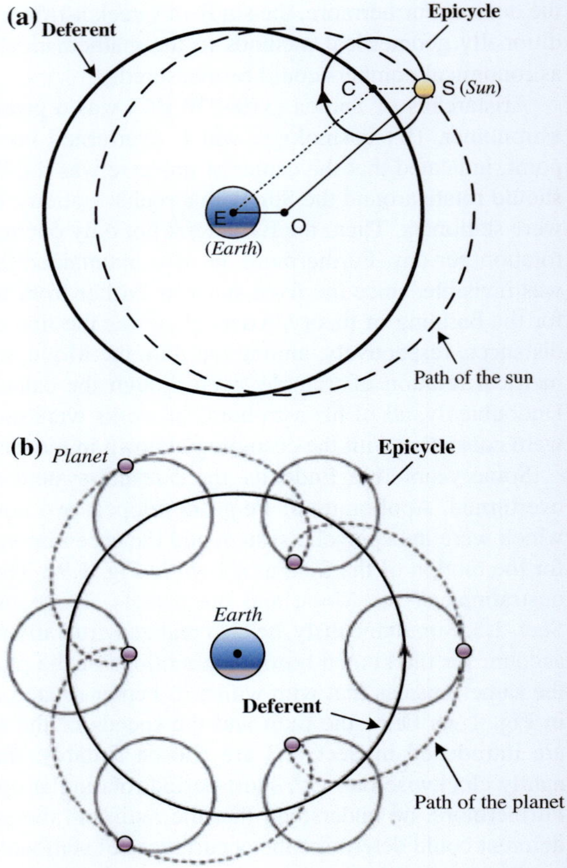

Fig. 1.9 Apollonius' equivalent astronomical models [28]. **a** Motion of the Sun. **b** Motion of the planet

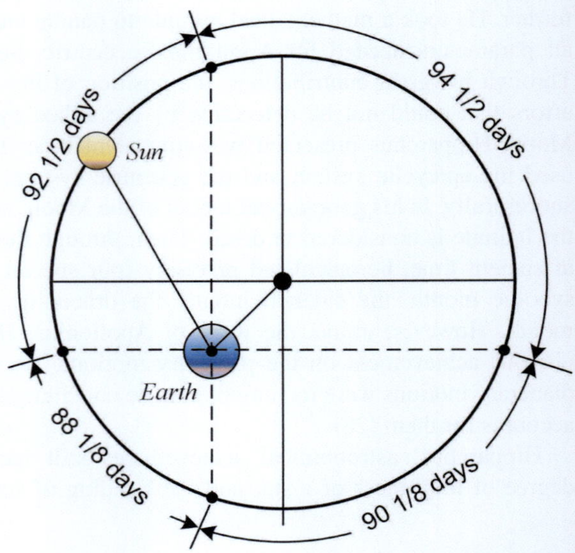

Fig. 1.10 Hipparchus' scheme for motion of the Sun [28]

observation, he realized that there were still several unknown stars in Heaven. Then, he built the star table with several newly discovered stars and distinguished the brightness of these stars by six degrees for the understanding of astronomical phenomena. The study of the stars never appeared in predecessors' works.

1.2 Astronomical Cycles and Calendars

There were numerous developed calendars and known astronomical cycles in various civilizations. They were referred to as the time standards at that time and were the outstanding achievement of astronomical observations. The intention here is not to introduce these ancient calendars and astronomical cycles one by one. In what follows, only the Egyptian calendar, the Metonic cycle, the Callippic cycle, the Saros cycle, and the Exeligmos cycle are introduced. These cycles and calendars are more related to the information emphasized in this book, and they are used in the later contents.

1.2.1 Egyptian Calendar

Calendars are the systems of time that include days, months, years, and some specific periods defined by mankind for agricultural, social, religious, commercial, or administrative purposes. They are established according to the observational records of various astronomical phenomena. For examples, a solar calendar is based on the apparent motion of the Sun; a lunar calendar is synchronized to the motion of the Moon; a lunisolar calendar is based on a combination of the solar and lunar motions; and some particular calendars appear to be synchronized to the motions of other celestial bodies, such as Venus and Sirius.

In early times, the Egyptians observed that once Sirius appeared in the Eastern sky at sunrise, the Nile River was again flooded. Then, they built the connection between the beginning of year and the cycle of Sirius. In their calendar, a year was 365 days long, which was divided into 12 months with 30 days and 5 extra days in the end of year. In accordance with the changes of seasons, an Egyptian year was also divided into 3 seasons, including the season of *akht* (inundation), the season of *peret* (winter), and the season of *shemu* (summer) [1, 4–6].

It is known in modern times that the length of a year is slightly more than 365 days. Since the Sirius took one revolution of 1460 years to return to the same position, the difference between a civil year and a true year was 365 days in 1460 years, i.e., one day in 4 years. This is called the Sirius cycle in the Egyptian calendar. In fact, the Egyptians understood the difference of time length of year, a quarter day, between the true time and their calendar. Therefore, they defined a rule of leap year in the calendar to reduce this time difference.

1.2.2 Metonic Cycle

An artificial calendar is generally represented in integral days. But the cycle of astronomical phenomena for defining a time standard is only an approximation and not an integral value. The error between the time length of calendar and the actual time length of astronomical phenomena must be inevasible but small. For a long time, the resulted error will be accumulated to disturb the calendars and their corresponding astronomical phenomena. Thus, some special means were executed for correcting the calendar synchronized to its astronomical phenomenon, such as the leap day of a solar calendar. Furthermore, there was another error resulting from the combination of two astronomical phenomena in a calendar. This means that a calendar was composed of two artificial time standards that were defined based on two cyclic astronomical phenomena, but the actual time lengths of these two time standards were different in a cycle of calendar. A lunisolar calendar is one of the classic examples. The lunar month was a time standard of the calendar defined in accordance with the observable period of the moon phase. In addition, the solar year was a time standard defined by the periodic motion of the Sun. A year of the calendar had 12 months, but the time length of 12 months (a year of calendar) was not consistent with the time length of the solar year. This made the calendar unable to satisfy the astronomical phenomena of the Sun to indicate the seasons. Therefore, astronomers used some means to eliminate such an error of the calendar so as to remain the consistence of true time and applied time standards.

Meton of Athens, who lived in the fifth-century BC, was a well-known Greek astronomer. He made an accurate observation of the astronomical phenomenon and defined the true length of a year much precisely. For his great works, the best known one is the calculation of the Metonic cycle that is a common multiple of the tropical year and the synodic month to adjust the consistence of the true lengths in days. Based on Meton's precise observations, it was confirmed that the length of 19 tropical years was almost exactly equal to that of 235 synodic months. Rounding to full days, the length of 19 tropical years counts 6940 days. Since 19 years only consisted 228 month in the ancient lunisolar calendar, 7 spare months must be added into the periods of 19 years. Therefore, one more month is traditionally added to the years 3, 6, 8, 11, 14, 17, and 19, i.e., the long years with 13 months [5].

1.2.3 Callippic Cycle

In the studies of astronomy and calendar, the Callippic cycle is an alternative approximate common multiple of the tropical year and the lunar month. It was the improvement of the Metonic cycle and derived by Callippus of Cyzicus in 330 BC. Meton's discovery, the cycle of 19 years that counted 6940 days, means that the length of a year is $6940/19 = 365 + 1/4 + 1/76$ days. But based on

Callippus' viewpoint, the length of a year is much bigger to generate the increasing difference between the tropical year and the lunar month. Furthermore, the length of a year should be closer to 365 + 1/4 days. Therefore, in order to eliminate the error of calculation, he built the 76-year Callippic cycle, which is 4 times of the Metonic cycle, to reach an integer number of days, and then dropped one day from the calendar after the forth Metonic cycle [5].

1.2.4 Saros Cycle

Astronomical cycles are the cycles of heavenly bodies' motions or celestial events that are interesting astronomical phenomena including one or more celestial bodies. In ancient times, several astronomical events that were visible to naked eyes, such as eclipse, lunar phase, conjunction, opposition, and retrograde and prograde motions, were understood. The astronomical cycles of some astronomical events were developed based on their observational records. A cycle of eclipse was a classic instance recorded in several civilizations such as ancient Greek, Babylon, Egypt, and China [7, 36, 37].

The Saros cycle, which is approximately 6585.3213 days or nearly 18 years 11 and 1/3 days, was able to predict the time of eclipse and calculated by the Chaldeans in the last few centuries BC [5, 37, 38]. It is an important finding with the effect of myth in the ancient astronomy, owing to the undeveloped knowledge. Its origin should be based on the recognitions of the synodic month, the anomalistic month, and the draconic month.

In ancient astronomy, it was known that the eclipse of the Sun or the Moon happened only when the Moon was new or full. Since the orbit of the Moon did not coincide with the common plane with the Sun, solar or lunar eclipses do not happen at every new or full moon in a month. Therefore, for an eclipse to happen, it is necessary that the Sun, the Moon, and the Earth are in a straight line and through the node of orbits. Furthermore, if the appearing times and the moon phases of two eclipses are the same, the distance between the Earth and the Moon should be the same for these eclipses. Based on above requirements of eclipses, the common multiple of synodic month, draconic month, and anomalistic month could be defined as the cycle of eclipse. Then, the Saros cycle is one approximate value that is equal to 223 synodic months, to 242 draconic months, and to 239 anomalistic months [5, 37, 38].

1.2.5 Exeligmos Cycle

The Saros cycle was an approximate value according to the above descriptions. Its time period was not an integer of days, but contained a multiple of 1/3 of a day. This factor made the complication of the eclipse prediction that should occur about 8 h later in the day. Nevertheless, even though the using of the Saros cycle was not convenient, the non-integral part of its time period could not be ignored. In order to improve the calculation of eclipse prediction, therefore, another cycle that is beneficial to predict successive eclipse was delivered. This astronomical cycle is the Exeligmos cycle [5, 39].

The Exeligmos cycle was the triple of Saros cycle, which approximated 19,756 days or 54 years plus one month. It is a better approximation of the cycle of successive eclipses and with the advantage that it has nearly an integer number of days. What's more, the Exeligmos cycle could provide a better succinct method. Its prediction result should be consistent with the result of the Saros cycle.

1.3 Ancient Astronomical Theories

The historical development of building the cosmology that could demonstrate the motions of celestial bodies was regarded as the emphasis of the ancient astronomy. Before the most known Ptolemaic system (Geocentric model) of the sixteenth century, the geocentric model had been the core of the cosmology in astronomy. In accordance with this cosmology, several astronomical theories and their corresponding geometrical models were successively developed, including the solar theory, the lunar theory, and the planetary theory. These astronomical theories depended on the astronomical developments in their eras. Although many historical manuscripts with these great works were lost, it is lucky to find the relevant historical records and descriptions in Ptolemy's "Almagest."

1.3.1 Solar Theory

In ancient times, the solar theory referred to the description of the solar motion with the viewpoint of mathematics. The resolution was usually a representation of geometric mathematics. As shown in Fig. 1.9a, Apollonius of Perga proposed two equivalent systems, including the epicyclic system and the eccentric system, to describe the solar anomaly motion [8, 28, 29]. Afterward, such a concept for describing the solar motion was proven and calculated by Hipparchus further, as shown in Fig. 1.10. This concept could be regarded as the major opinion of the solar motion during the time period of the Hellenistic Age. It was the solar theory at that time.

In the epicyclic system, there are two circles, called the epicycle and the deferent, with the Earth located at the center E of the deferent, and the Sun S positioned on the epicycle. Center C of the epicycle rotates about the Earth around the deferent. The epicycle rotates in the opposite direction to which the center of the epicycle rotates. All of the rotations in this system should be at the same angular speeds, i.e., once every tropical year. Alternatively, the solar motion is considered in the eccentric circle system. It was around the dashed circle of which center O

called the eccenter. The position of the Earth is away from the eccenter with the distance known as eccentricity. In summary, the above solar theory represented a solar anomaly motion but with a periodic variation of its rate.

1.3.2 Lunar Theory

The lunar theory means the description of the lunar motion with an analytical and mathematical concept. As mentioned above, the lunar motion had been understood as the anomaly motion around the Earth during the time period of the Hellenistic Age. It is evidenced that Hipparchus developed two equivalent lunar theories [1, 28, 29, 40]. These two theories corresponded to various geometric models including the epicyclic systems and the eccentric system mentioned in the solar theory.

The first theory indicated that the Moon rotates at the rate of the sidereal month about an eccenter, which in turn rotates about the Earth at the rate of the anomalistic month. The second one indicated that the Moon rotates on an epicycle at the rate of the anomalistic month relative to a deferent circle that rotates at the rate of the sidereal month. Owing to the confirmed equivalence of these two systems, only the epicyclic system was analyzed to obtain the resulting kinematic equation of the lunar motion. The relevant contents are introduced in Chap. 7.

1.3.3 Planetary Theory

The planetary theory refers to the description of various planetary motions with analytical and mathematical concepts. Apollonius successfully explained the solar and lunar motions by his geometrical models. In the same time, he used the epicyclic systems to simulate the known planetary motions. Figure 1.9b shows the imaginative path of a planetary motion according to Apollonius' opinion. The epicycle and the deferent must have opposite rotational directions and various rotational rates [28, 29]. Thus, the retrograde motion could be explained by this geometrical model.

The rates (periods) and the radii were essential for the geometrical model of planetary motions. These data should be concluded based on the observation records of the corresponding astronomical events. Ptolemy's "Almagest" clearly expounded the required data of the geometrical model of the five planets, but there are no existing historical records relevant to them in the Hellenistic Age [28, 29, 41]. This situation only makes it difficult for use to rebuild Apollonius' geometrical model of planetary motions. Since it is evidenced that they were already understood and obtained according to the much earlier Babylonian astronomical records, it is logical to deduce that the data of planetary motions had been resolved in the Hellenistic Age. And thus, these data could affect the Ptolemy's data or be described in the manuscript of the "Almagest."

1.4 Remarks

Owing to the close geographical positions, the historical developments of Greek astronomy were affected by the Mesopotamian and the Egyptian civilizations. The observation records of the astronomical phenomena and events, the astronomical theories, and the cosmologies of these civilizations were inherited and developed further in Greece. Certainly, the calendars built in different eras were also adopted with slight corrections.

Aegean civilization is the earliest one in Europe. Its beginning can be traced back to the Minoan and Mycenaean civilizations, and the relevant actions of astronomy were developed in that time. It was proven by the excavation, a Minoan clay disk with a calendar record. Form the Dark Age to the Hellenistic Age, several schools with different cosmologies were developed in turns. The geocentric model, the Earth was the center of universe, was built completely. Then, the geometrical models to demonstrate motions of the celestial bodies were claimed.

One of the great achievements of ancient astronomy is the building of various calendars based on long-term observations and records of the cyclic astronomical phenomena. The defined time systems were contributed to the society, the administration, and the livelihood. Some special ways used to eliminate the error between the actual time and the artificial calendar were presented. In addition, the Egyptian calendar, the Metonic cycle, the Callippic cycle, the Saros cycle, and the Exeligmos cycle were introduced.

Finally, the astronomical theories for the solar, the lunar, and the planetary motions in the Hellenistic Age are introduced, i.e., two developed geometrical models describe these motions: the epicyclic system and the eccentric circle system. These geometrical models could not only demonstrate the motions of celestial bodies but also explain their particular astronomical phenomena. It is concluded that the motions of these celestial bodies are anomaly motions.

References

1. de Vaucouleurs G (1957) Discovery of the Universe. The Macmillan Company, London
2. de Vaucouleurs G (2010) A brief history of astronomy. Chinese Edition: G.de Vaucouleurs (2010) Tian Wen Xue Jian Shi (trans: Heng Li), Zhongguo ren min da xue chu ban she (in Chinese), Beijing
3. Clagett M (1995) Calendars, clocks, and astronomy, ancient Egyptian science, 2nd edn. American Philosophical Society, Philadelphia
4. Parise F (1989) The book of calendars. Facts on File Inc., New York
5. Zheng Y, Yu ZK (1995) Shi Jian Yu Li Fa (in Chinese). Yin Hegu, Taiwan
6. Database of Greek Cosmology (2003) Chapter 2: Greek Cosmology. In: Notes from the class of Physics 7: Relativity, Space-Time and Cosmology. University of California, Riverside, California, USA. Available in the website. http://www.physics.umd.edu/courses/Phys117/Liberati/Support/cosmology.pdf. Accessed July 2014
7. Hoskin M (1997) The Cambridge illustrated history of astronomy. Cambridge University Press, Cambridge

8. Kenton E (1928) The book of Earths: the Egyptian universe. (Online edition). Forgotten Books. http://www.forgottenbooks.org. Accessed July 2014
9. Wilson R (1997) Astronomy through the age: the story of the human attempt to understand the Universe. Taylor & Francis, London
10. Excavation of Dendera temple (50 BC) The collection of Louvre Museum, Paris. Available in the Museum website. http://www.louvre.fr/en/oeuvre-notices/zodiac-dendera. Accessed July 2014
11. Ancient Wisdom and Modern Day Confusion (2009) Emergent Culture. Available in the website. http://emergent-culture.com/preview-relating-web-bot-synchronicity-mayan-calendar-the-dna-and-future-forecasting/comment-page-1/. Accessed July 2014
12. Ibrahem A (2014) Egyptian cosmology. In: Eclipse chasers. Available in the website. http://www.eclipse-chasers.com/akhet.html. Accessed July 2014
13. Pillars of Heaven (2014) Institute of biblical and scientific studies. Available in the website. http://www.bibleandscience.com/bible/books/genesis/genesis1_pillarsheaven.htm. Accessed July 2014
14. Walton J (2012) Genesis through ancient eyes: part 1 (video). The Biologos Forum-Science and Face in Dialogue. Available in the website.http://biologos.org/blog/genesis-through-ancient-eyes-part-1. Accessed July 2014. T. Larsen (eds) (2013) Ancient cosmologies and the creation story of genius. Available in the website. http://relevancy22.blogspot.tw/2013/03/how-to-interpret-creation-story-of.html. Accessed July 2014
15. Coming Forth by Day at Ares Vallis (1997) SiLoamNet. Available in the website. http://www.siloam.net/members.aol.com/rushjudi/private/mrs001.htm. Accessed July 2014
16. Stray G (2007) The Mayan and other ancient calendars. Walker Publishing Company Inc., New York
17. Heidel A (1942) The Babylonian Genius: the story of creation. The University of Chicago Press, London
18. Warren WF (1915) The Universe as pictured in Milton's paradise lost. The Abingdon Press, New York. Available in the website: University of Toronto Libraries. https://archive.org/details/universepictures00warruoft. Accessed July 2014
19. Radau H (1902) The creation story of genius I: Sumerian theogony and cosmogony. The Open Court Publishing Company, Chicago. Available in the website: American Libraries. https://archive.org/details/creationstoryge00radagoog. Accessed July 2014
20. Knowledge of Babylonian Tablets (2013) Babylonian Culture and Tablets. Available in the website. http://www2.econ.iastate.edu/classes/econ355/choi/bab.htm. Accessed July 2014
21. Budge EAW (1921) The seven tablets and creation: description of their content. The Babylon Legend of Creation. Available in the website. http://www.sacred-texts.com/ane/blc/blc07.htm. Accessed July 2014
22. Knowledge of Mesopotamia Cosmology (2010) The firmament of genesis 1 is solid but that's not the point. In: The Biologos Forum-Science and faith in dialogue. Available in the website. http://biologos.org/blog/the-firmament-of-genesis-1-is-solid-but-thats-not-the-point/. Accessed July 2014
23. Jones A (1991) The adaptation of Babylonian Methods in Greek numerical astronomy. ISIS 82(3):441–453
24. McGee M, Shanks M (2006) Ancient Greece: archaeology unlocks the secrets of Greece's past. National Geographic Investigates, Washington DC
25. Ridderstad M (2009) Evidence of Minoan astronomy and calendrical practices. Cornell University Library, ArXiv.org, pp 1–42
26. Henriksson G, Blomberg M (1996) Evidence for Minoan astronomical observations from the peak sanctuaries on Petsophas and Traostalos. Opusula Atheniensia 21(6):99–114
27. Williams B (2002) Calendars. Cherrytree Press, Berkshire
28. Linton CM (2004) From Eudoxus to Einstein: a history of mathematical astronomy. Cambridge University, New York
29. Dreyer JLE (1953) A History of Astronomy from Thales to Kepler. Dover Publications Inc., USA

30. Wudka J (1998) The pythagoreans. Available in the website. http://physics.ucr.edu/~wudka/Physics7/Notes_www/node32.html Accessed July 2014
31. Geocentric Model (2014) Universe. Reference library, RedOrbit. Available in the website. http://www.redorbit.com/education/reference_library/space_1/universe/2574692/geocentric_model/. Accessed July 2014
32. Motz L, Weaver JH (1989) Chapter 2: Greek astronomy. In: The story of physics, Springer, New York
33. Padersen O (1974) Early physics and astronomy: a historical introduction. Macdonald and Janes and American Elsevier Inc, New York. (Republication) (1993) Press Syndicate of the University of Cambridge, New York
34. Heath TL (1932) Greek astronomy. J. M. Dent & Sons, Ltd., London. (Republication) (1991) Dover Publications, New York
35. Mccrone J (2006) Anaximander. In: History. Dichotomistic. Available in the website. http://www.dichotomistic.com/logic_dichotomies_history_one.html. Accessed July 2014
36. Feng S (2005) Tian Wen Xue Shi Hua (in Chinese). Guo Jia, Taipei
37. Brady B (1995) The Saros Cycle—Eclipses come in Families. In: Bernadette Brady. Available in the website. http://www.bernadettebrady.com. Accessed July 2014
38. Bulgerin M (2002) Eclipses and the Saros Cycle. Available in the website. http://members.bitstream.net/~bunlion/bpi/EclSaros.html Accessed July 2014
39. Knowledge of Exeligmos (2011) Exeligmos: the Triple Saros cycle of eclipses. In: The astrology reading of a lifetime. Available in the website. http://www.myastrologybook.com/Exeligmos-Triple-Saros-eclipse-cycle.html. Accessed July 2014
40. Freeth T et al (2006) Supplementary notes. In: Decoding the Ancient Greek astronomical calculator known as the Antikythera Mechanism. Nature 444:587–891
41. Ptolemy C (100) Almagest. English edition: C. Ptolemy (1998) Ptolemy's Almagest (trans: G.J. Toomer). Princeton University Press, USA

Chapter 2
Ancient Astronomical Instruments

With the developments of ancient astronomy, ancient astronomical mechanical instruments were extensively used for measurements, computation, and demonstration. These instruments represented the combinational achievements of astronomy and mechanical engineering technology at the subjects' eras. The investigations of the recorded ancient astronomical instruments not only give us a glimpse of their historical developments, but also provide overall concepts and knowledge about the mechanical characteristics for some designated functions and applications. Furthermore, the relevant craftsmanship is realized. This chapter presents a study of the ancient astronomical mechanical instruments and classifies them in accordance with their display functions. Some representative artifacts such as sundials, astrolabes, calendrical device and relevant compound instruments, astronomical (mechanical) clocks, and orreries are introduced. Focusing on a specific display function, the designs among different astronomical instruments are compared. In addition, the manufacturing technology and craftsmanship for generating some particular functions are analyzed.

2.1 Classifications Based on Functions

Designs of mechanical devices are advanced with the historical developments of diverse cultures, sciences, and manufacturing techniques along with numerous applications. Astronomy is one area much affected in ancient cultures. The characteristic of constrained motions of mechanical devices is widely used for the astronomical applications to produce complicated calculations, precise measurements, and ingenious exhibition. The development of ancient astronomy attaches to several ancient astronomical instruments with sophisticated designs and special applications. For examples, sundials, astrariums, astrolabes, planetariums, and

J.-L. Lin and H.-S. Yan, *Decoding the Mechanisms of Antikythera Astronomical Device*, DOI 10.1007/978-3-662-48447-0_2

astronomical mechanical clocks are well-known devices with various knowledge of mechanisms and astronomy.

In what follows, the available ancient astronomical mechanical instruments, based on their functions, are classified into the applications of measuring time, measuring position, observation, computing, and demonstration [1–9].

2.1.1 Observation Application

These refer to the mechanical instruments used to assist in observing the phenomenon of celestial bodies in yonder universe. Observational astronomy is a division of modern astronomy. It is the practice of observing celestial objects by using telescopes and other astronomical apparatus. The observation of astronomical phenomenon is the leading work of ancient astronomy for recording and understanding what happened in the sky and their cycles by simple mechanical devices.

The changes of seasons are conscious natural phenomenon; however, it is closely constrained by the areas and climates the observers live in. Once the climate change of area is not obvious, the time cycle is not defined. Therefore, the revolutions of phenomenon of celestial bodies are regarded as the most popular and steady units of time. According to this requirement, the historical records about the astronomical observations existed in many ancient civilizations. Also, it is related to the developments of mechanical instruments for astronomical observation. Telescope is the representative work of observational astronomy. The first one was made in Holland around 1608 [5, 6]. In ancient Western civilizations, *cross-staffs* and *astrolabes* could be regarded as the instruments to assist the observation of astronomical phenomenon and events.

2.1.2 Measuring Position and Distance Application

These refer to the mechanical instruments for identifying positions of celestial bodies in the sky and the angles between each other, even to estimate the distances between these objects and the observers. Depending on the contribution of geometrical mathematics, the distance between measured objects could be estimated. In general, the application of measuring position or angle can be regarded as a part of astronomical observation work in ancient times.

The developments of measuring devices start with the study of geometry. For measuring distance, the measuring devices were usually used in the building of towers, temples, and towns [2, 6]. In addition, they can measure the distance with a wide range that was difficult to cross over such as a river, a mountain valley, or a sea inlet. For navigation, navigators used them to measure the elevation angle of the noontime Sun above the horizon, which allowed them to estimate their latitude. For astronomers, they were used to measure the angle between the directions of two celestial bodies or position heavenly bodies in the sky before the invention of telescopes. The aspiration was to estimate the distance between the Earth and stars. *Cross-staffs* and *astrolabes* are two popular ancient astronomical mechanical instruments not only for observation, but also for measuring.

2.1.3 Measuring Time Application

These refer to the mechanical instruments for counting or indicating some time standards, which did not have obvious phenomenon (such as sunrise, sunset, and moon phase) to distinguish their time intervals.

Measuring position (angle or distance), measuring time, and observation can represent the first step in developing the combination of astronomy, mechanical design, and science and technology [2]. The work of measuring time starts with the definitions of a day, a month, a year, and other time cycles. These time definitions were based on the long-term observational records of cyclic astronomical phenomenon. Sunrises and sunsets are the basic standards to distinguish a day. For Egyptian and Romans, they further divided their day into 24 h, like we do today. Among these 24 time intervals, 12 h are for the day and another 12 h are for the night [1, 3, 5, 6]. Even though this definition resulted that an hour of the day is not the same as an hour of the night in some seasons, an hour was contributed to distinguishing and indicating the time much clearly in ancient times. Hence, the mechanical devices for measuring time were designed to distinguish the hours of a day.

In ancient times, the hours of a day were identified based on the position of the Sun in the sky. For instance, *sundial* is the ancient measuring device that used the shadow of object, *gnomon*, to reflect the position of the Sun. In addition, the users could understand the time according to the time graduation that the shadow indicated. Nevertheless, the operation of the device is determined on the appearance of the Sun. It was useless on days when there was no Sun, and it could not measure the nighttime. The appearance of clocks overcomes the above constraints of the sundials.

Clocks were an important invention for measuring time. The precursor of clocks was the water clock [1, 3]. It depended on the flow of water from a tank to indicate the hours, as shown in Fig. 2.1 [10]. Because of the different duration of hours of the day and the night, the tank could be designed to coincide with the duration of the day by varying the quantity of a scale. Afterward, the *mechanical clock* was invented based on the redefinition of hours, and its service depended on the accurate motions by gearing. Many early mechanical clocks were not a separate instrument for measuring time. They were *astronomical clocks* for measuring time, displaying astronomical motions, and showing astronomical information. Such a sophisticated device was regarded as a combination of outstanding manufacturing technique and advanced astronomical achievement. It could continuously demonstrate the cyclic astronomical motions according to the clockwork.

Fig. 2.1 Virtual imitation of ancient Egyptian water clock

2.1.4 Computing Application

The astronomical mechanical instruments for computing applications refer to the devices with mathematical gearings for calculating the astronomical cycles. For mechanical devices, the power output or transmission was not the only required application. This type of instruments focused on using the kinematic characteristics of mechanisms to generate specific rate of transformations. Generally, such an output function was achieved by gear mechanisms. The resulting mechanical instruments could calculate some astronomical cycles or calendars based on the known mechanism characteristics. In addition, these astronomical mechanical devices seemed to be the precursor of computation devices [4].

In fact, the ancient astronomical mechanical devices with computing applications were used for calendrical devices. According to available historical records, they were mostly not formed as independent instruments, but usually associated with other astronomical instruments, such as *sundials* or *astrolabes*.

2.1.5 Demonstration Application

The astronomical mechanical instruments for demonstration application are defined as the devices used to simulate relevant academic and science theories by mechanical designs, such as planetariums, *armillary spheres* (*concentric spheres*), and *orreries*. Simultaneously, their displays could be regarded as the computation results to show the attitude, latitude, and/or longitude of designated positions.

As mentioned in Chap. 1, it is understood that the ancient astronomy had developed the hypothesized geometrical models of the universe and the orbits for describing the motions of heavenly bodies based on the observational records.

These geometrical models in writing form were applied on the physical mechanical devices that generated the motions satisfied their corresponding astronomical geometrical models, owing to the development of manufacturing technique and the advanced understanding of mechanism characteristics. Some complete mechanical devices referred to the embryo of the universe that was opinioned at that time. Like the astronomical computing devices, they sufficiently applied the kinematic characteristics of mechanisms for generating appropriate rate ratios and built by gear mechanisms. Furthermore, they were the essence of astronomy in their eras.

2.2 Jacob's Staff

Jacob's staff was a widely used ancient measuring device. It, also called *baculum*, *cross-staff*, or *radius*, consists of a main staff with a perpendicular crosspiece, which was attached to the staff at its middle and was able to slide up and down along the staff, as shown in Fig. 2.2 [2, 5, 8, 9, 11–13].

In navigation, this apparatus was used to determine angles or distances: The angle between the horizon and Polaris (or the Sun) can be determined to estimate the latitude at which a vessel is located. The angle between the top and bottom of an object can be measured to determine the distance to the object, based on geometric mathematics, if its height is known, or to determine the height if its distance is known. Furthermore, the horizontal angle between two positions can be measured to determine the other position on a map [2, 9, 11, 12].

When Jacob's staff was used for astronomical observation, it was referred to as a *radius astronomicus*. Figure 2.3 shows a schematic introduction of application method of a staff. In order to measure the angle between two stars or the altitude of a star, astronomers would place the staff below one of his eyes. Since the crosspiece would have a pair of open sights (sticking out perpendicular to the scheme)

Fig. 2.2 Virtual imitation of Jacob's staff

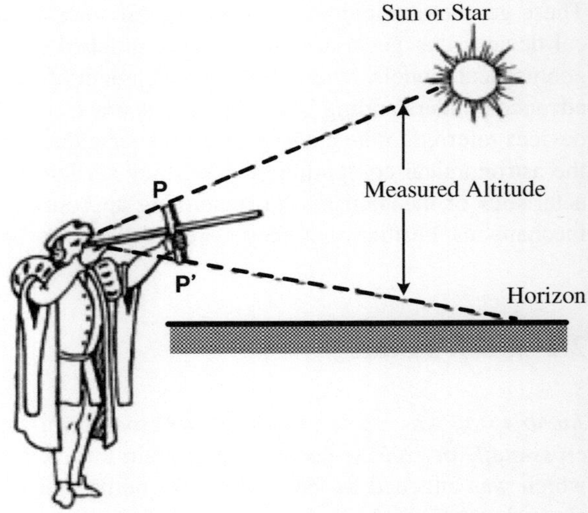

Fig. 2.3 Application illustration of Jacob’s staff [11]

at symmetric locations such as P and P′, they slide the crosspiece up and down until sight P covered one of the stars and sight P′ the other. Moreover, Jacob’s staff was workable at night because of the convenient slits on the sights [9, 11, 12].

2.3 Astrolabe

An *astrolabe* is an astronomical instrument used for measuring altitudes of celestial bodies, displaying the positions of stars, and finding the time of a day and the time of celestial events, such as sunrise and sunset [14]. Its origin was in the classical Greece (lasting from the fifth through the fourth century BC). The astrolabe was introduced to the Islamic world until the mid-eighth century and was highly developed and extensively spread during the early centuries of Islam. With the popularization of Islam, it moved through North Africa into northern Spain and was introduced into Europe in around eleventh century [15–19].

The stereographic projection is the most important theory for building an astrolabe. It was as if Apollonius had studied the theory. Afterward, Hipparchus refined the structure of the theory. He redefined and formalized the concept of projection to build a method for solving complex astronomical problems. Hipparchus is regarded as the most influential individual in the discovery of the stereographic projection theory, but he did not invent the astrolabe [11, 16]. The earliest evidence for applying the stereographic projection theory in a mechanical device was an anaphoric clock in around the first century. Nevertheless, exactly when the stereographic projection theory was turned into the astrolabe, we known today remains a mystery.

A traditional astrolabe was composed of several members, as shown in Fig. 2.4 [15]. Its major members were rete, mater, and plate. A rete is a skeletal plate that

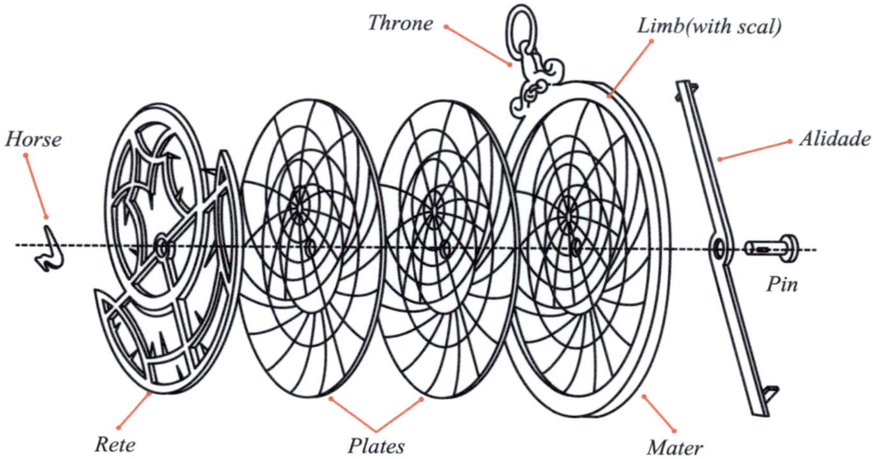

Fig. 2.4 Members of astrolabes; reprinted from Ref. [15]

fits into an astrolabe on top of the selected latitude plate and is freely rotatable over it. The rete is the stereographic projection of celestial sphere. It had star pointers and the stereographic projection of the ecliptic. The star pointers were placed at the positions of their stereographic projection to mark visible and bright stars [14, 15, 18, 19]. The form of the pointers could be simple indicators or patterns of vines and animals. The retia of the astrolabes are various. For early astrolabes, the number of star pointers varied from 10 or 12, and they increased to as many as 50 or 60 on later ones. They were regarded as the representative symbol of a family. Besides, a mater is the disk of an astrolabe which usually had a hollow section. A plate for representing the local latitude engraved with circles of altitude and azimuth for a certain latitude was inserted into the hollow section of the mater. It is the stereographic projection of the local sky based on the latitudes of builders' positions. Sometimes, the ancient astrolabes included several plates engraved on both sides, so the instrument could be used in various latitudes. For the astrolabe, the projection of the celestial sphere and the projection of the local sky are overlaid with a common point, the north celestial pole. Thus, the projection of the celestial sphere can be rotated with respect to the projection of the local sky to display a view of the sky for any combination of date or time [15, 19].

Whenever an astrolabe was used, the moveable members were adjusted to a specific date and time. After appropriately setting, much of the sky was represented on the instrument. The two faces of the astrolabe had their own functions. Most problems were solved by the front of the astrolabe which has fixed and rotatable parts. The fixed parts represent timescales and the stereographic projection of the sky. The rotatable parts were adjusted to simulate the daily rotation of the sky. As to the back of the astrolabe, it usually had scales for finding the longitude and the altitude of the Sun or a star for a specific date.

2.4 Sundial

A *sundial* is a device for measuring the hour of a day existed in many civilizations in the ancient world. The oldest sundials appeared around 3500 BC and were represented by obelisks and solar clocks by Egyptians and Babylonians [3]. In general, a sundial consists of the *gnomon* that is also called the *style* and the surface engraved with the timescale for indicating the time of a day, as shown in Fig. 2.5. The shadow of the gnomon is projected on the surface. When the Sun moves in the sky, the corresponding gnomon's shadow moves in the surface. According to the position of the gnomon's shadow, the hour of a day is identified.

The shape of gnomon might be that of a road or an edge, which was not constrained, and the styles of the surface were diverse. Ancient sundials could be

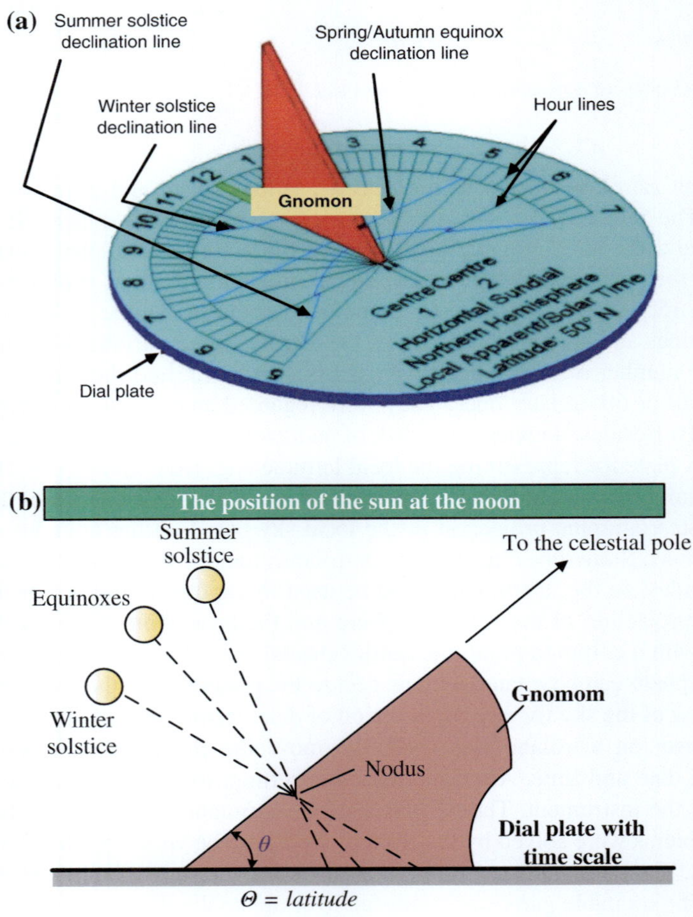

Fig. 2.5 Illustration of the sundial. **a** Members. Reprinted from Ref. [40], copyright 2015, with permission from Carl Sabanski. **b** Projection [41]

classified into two types: fixed sundials and mobile portable sundials. Fixed sundials refer to sundials that are mounted in a fixed structure. According to the styles of the sundials' surface, fixed sundials can be further divided into equatorial sundials, horizontal sundials, vertical sundials, and non-planar sundials, as shown in Fig. 2.6. The surface of equatorial sundials is an incline parallel to the equator, and its gnomon is parallel to the axis of the Earth. The surface of horizontal sundials is parallel to the ground, and the gnomon is parallel to the axis of the Earth. These

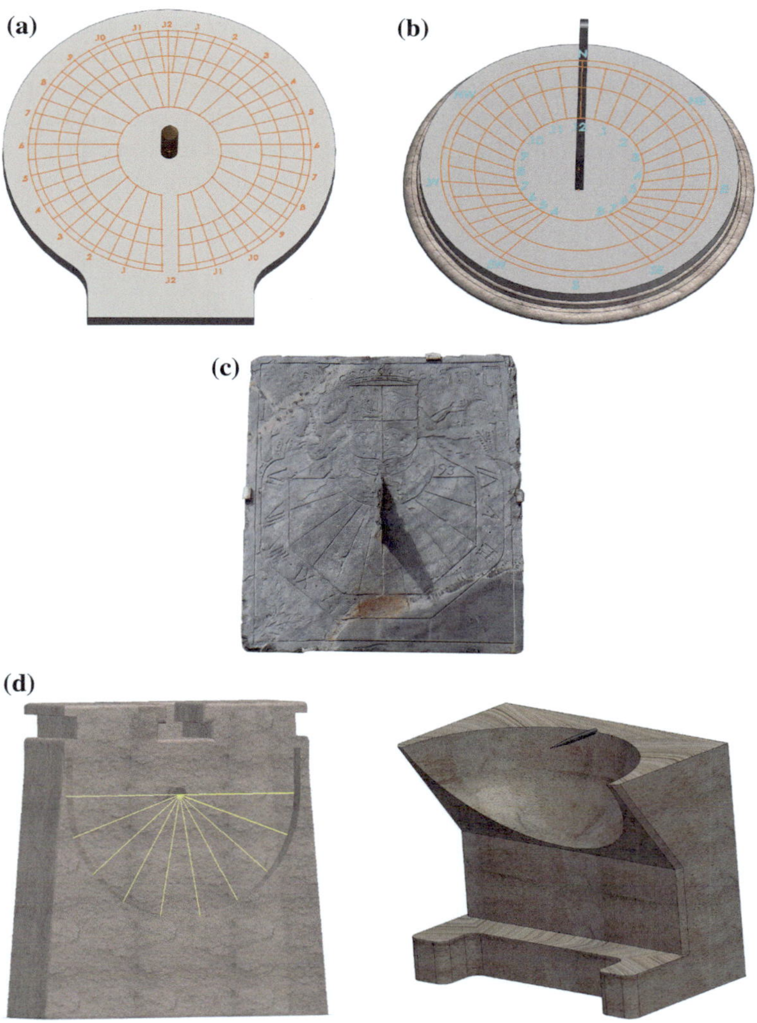

Fig. 2.6 Fixed sundials. **a** Equatorial sundial [42]. **b** Horizontal sundial. **c** Vertical sundial, reprinted from Ref. [43], copyright 2015, with permission from the Web site. **d** Non-planar sundial [44, 45]

two types of sundials were much popular in the excavated sundials. The surface of vertical sundials must be vertical to the ground, and their gnomon is still aligned with the axis of the Earth. As a result, this type of sundials was usually built on the wall. The final type is non-planar sundials with curved surface, like a part of inter-surface of a cylinder, a cone, or a sphere, to receive the shadow of the gnomon [3].

There were many styles of portable sundials developed in ancient times. The excavated fragment of an Egyptian portable sundial for example [19], in general, this type of sundials still had a gnomon that was a perpendicular block rising at the foot of the sloping face, as shown in Fig. 2.7. The height and width of the gnomon and the sloping object were the same. Whenever the instrument was to be used, it must be put down on a flat surface and be turned so that it faced the Sun directly [20, 21]. Thus, the shadow of the gnomon could fall upon the sloping face for showing the time. Furthermore, there were two special types of portable sundials, except for the surviving Egyptian portable sundial. One is a diptych sundial, as shown in Fig. 2.8. This type of portable sundials had two leaves. For a simple

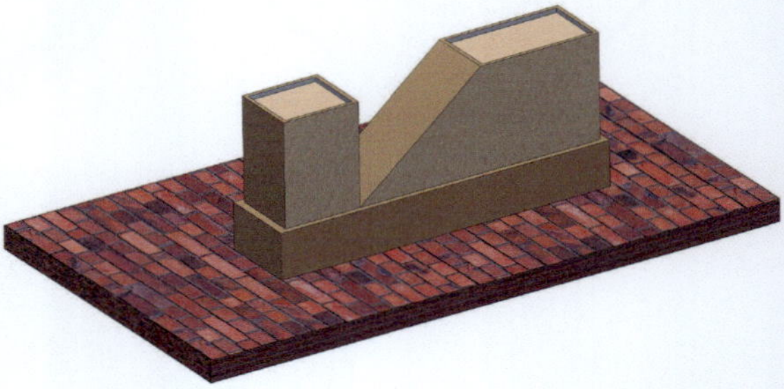

Fig. 2.7 Virtual imitation of Egyptian portable sundial [21]

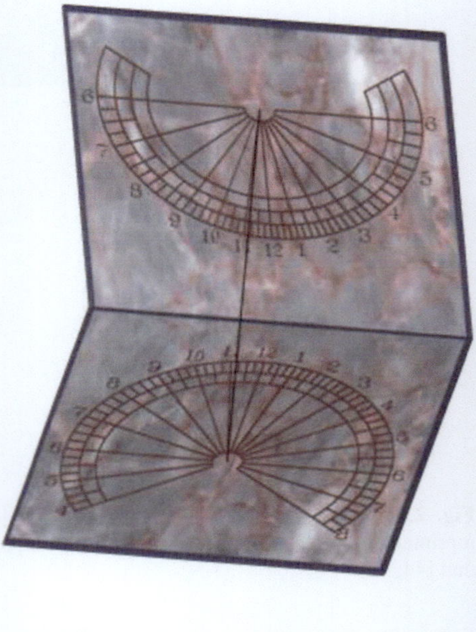

Fig. 2.8 A simple diptych (portable) sundials [40]

Fig. 2.9 Ring sundial [43]

diptych dial, the inter-face of one leaf was a vertical sundial and the inter-face of the other leaf was a horizontal sundial. Such a sundial can be adjusted according to users' latitude, by inclining it until its gnomon is parallel to the axis of the Earth. Actually, most portable sundials included a compass contained for orientation and are served completely in museums at present [19]. The other is the ring sundial, as shown in Fig. 2.9. Its main body was a ring held by a cord and had a hole through which lights could pass. Thus, the resulting light point could project on the ring with a timescale to indicate the hour of a day.

2.5 Calendrical Device

It is well known that one of the characteristics of gear mechanisms is to transmit motions with designated gear ratios. According to finding of archaeology, such a technique was used widely in the building of calendrical devices [22–28]. As mentioned above, all known ancient calendrical devices were associated with other astronomical mechanical devices to form multifunction astronomical mechanical devices until now. In what follows, some of these special mechanical devices are introduced.

2.5.1 Astrolabe with Calendrical Gearing

The astrolabe with calendrical gearing is the combination of the calendrical device and the astrolabe. It is an ancient astronomical mechanical device with compound functions. Except for the fundamental function to understand the night sky at some observation position, this device can also operate the calculations of calendars through the application of mathematical gearing.

A complete historical record of mathematical gearing, as shown in Fig. 2.10a, illustrated the calendrical device by Al-Bīrūnī (973–1047), who is regarded as one of the greatest scholars of the medieval Islamic era and was versed in physics, mathematics, astronomy, and natural sciences [22–26]. Since relevant historical records did not definitely describe what mechanical devices contained such a calendrical gearing, the definition of the calendrical gearing is uncertain. It is mainly believed that the gearing could be associated with an Islamic astrolabe based on the investigation of other similar mechanical devices. But, it is believed that the gearing was a separate mechanical device and was a rare ancient astronomical device. This gearing is described completely in the manuscript of the Islamic treatise in 1000. And it functions to model the cyclic variation of moon phases and the calendars. Figure 2.10b illustrates Al-Bīrūnī's relevant astronomical work for explaining various moon phases. Furthermore, Fig. 2.10c is the mechanical scheme of this mathematical gearing in the manuscript. It was clearly indicated that the mechanism was an ordinary gear train consisting of 8 gears and 5 separate arbors. Through the analysis of gear ratios, the gearing is known as using axis *B* as the input, that rotates once every 7 days; axis *A* rotates once every 28 days to display the position of the Moon; axis *C* rotates once every 59 days to reveal the age of the Moon; and axis *E* rotates once every 366.42 days.

Another example is the Persian astrolabe with calendrical gearing that dated back to AD 1221–22 and has been preserved in the Museum of the History of Science of Oxford [23, 24, 29], as shown in Fig. 2.11a. The front plate is for astrolabe, and two lugs on the rete serve as sights. The back plate is for the calendar without the use of a normal alidade. This device was operated by turning the central pivot, probably by hands or by using the astrolabe rete as a handle. The corresponding calendar and the lunar phase are moved accordingly. This compound mechanical device is found to consist of the gearing with the same purpose as that in Islamic treatise but with different gear ratios. Figure 2.11b shows the mechanical scheme of the calendrical gearing of this device. This gear train consists of 7 gears and 5 separate arbors. According to the analysis of gear ratios, the functions of this calendrical gearing are known as using axis *B* as the input, that rotates once every 7.375 days; axis *A* rotates once every 27.23 days to display the position of the Moon; axis *C* rotates once every 59 days to reveal the age of the Moon; and axis *E* rotates once every 354 days.

2.5.2 Sundial with Calendrical Gearing

This type of mechanical devices is the combination of the sundial and the calendrical device. A known example is the Byzantine sundial with calendrical gearing dated from late-fifth century of the Christian era or the first half of the

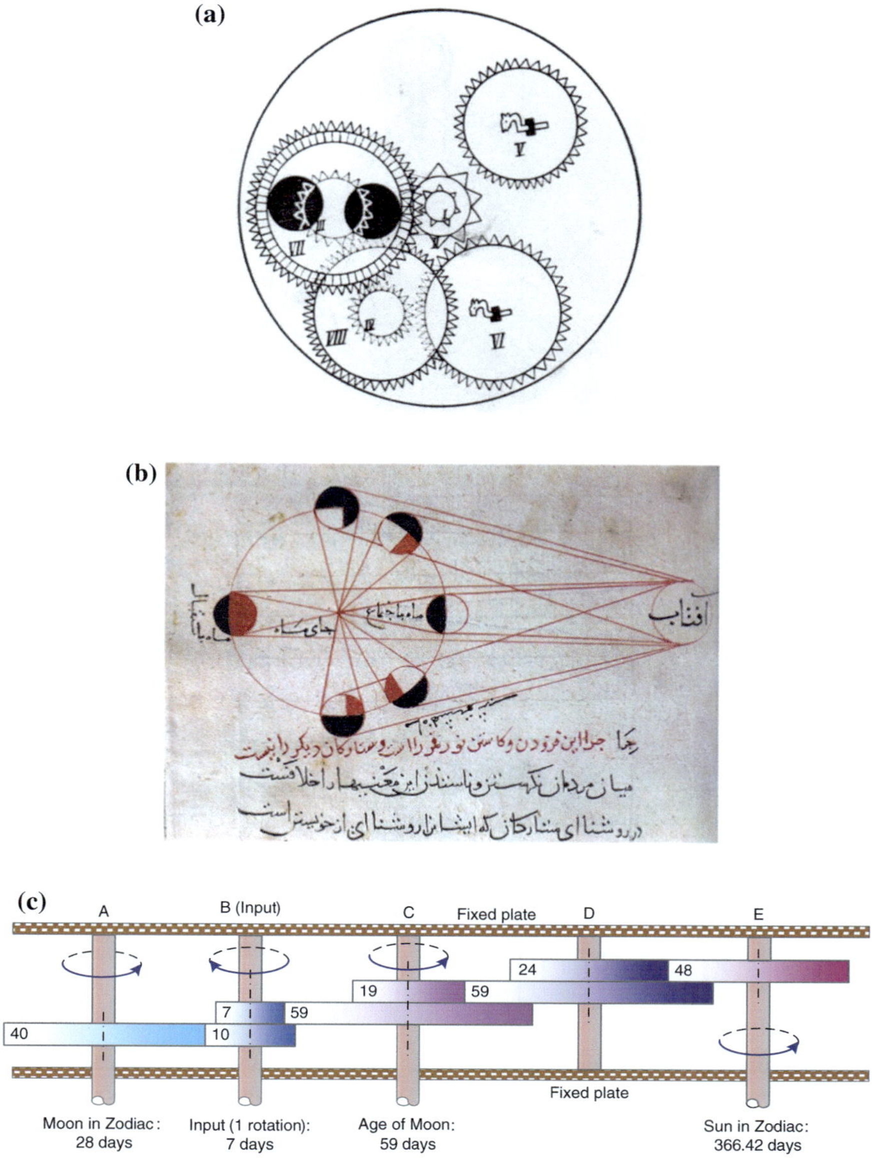

Fig. 2.10 Calendrical device by Al-Bīrūnī. **a** Historical picture, reprinted from Ref. [24]. **b** Illustration of moon phase, reprinted from Ref. [43]. **c** Mechanical drawing [23]

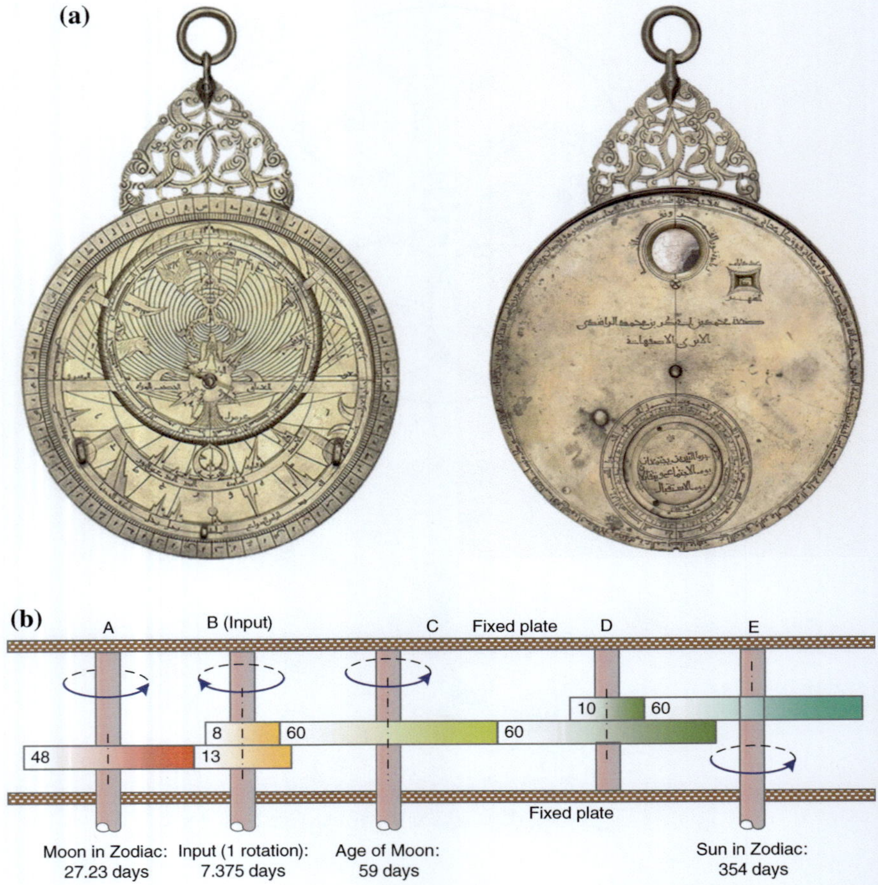

Fig. 2.11 Persian astrolabe with calendrical gearing. **a** Antiquity, reprinted from Refs. [14, 19, 24], copyright 2015, with permission from Museum of History Science. **b** Mechanical drawing [23]

sixth century [23–29]. It was much earlier than the Islamic calendrical device. The excavation of this mechanical device survives as fragments including 2 visible arbors [23, 30]. In 1984, a reconstructed model of the Byzantine sundial with calendrical gearing was presented by J.V. Field and M.T. Wright, as shown in Fig. 2.12[23, 30]. The calendrical gearing of the reconstructed model adopted similar design with the historical record of Islamic treatise and the Byzantine sundial with calendrical gearing to display the positions of the Sun and the Moon.

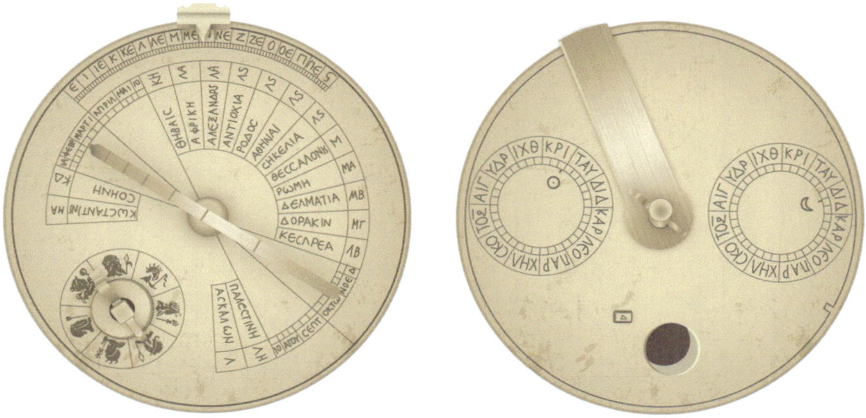

Fig. 2.12 Reconstruction model of the Byzantine sundial [23, 30]

2.6 Planetarium, Astrarium, and Astronomical Clock

Planetariums are diverse with ages. In early times, the planetarium was referred to as a mechanical device that could demonstrate the cyclic motions of heavenly bodies based on the universe model proposed at that time. This mechanical device was driven by gear mechanisms. However, in modern times, a planetarium usually refers to a theater built primarily for presenting educational and entertaining shows about astronomy and the night sky, or for training in celestial navigation.

An *astrarium* was also defined as a mechanical device for displaying the motions of the celestial bodies. It was generally for the *astronomical clock* [23]. According to historical literatures, this type of astronomical instruments is an outstanding achievement in the fourteenth century. The earliest astronomical clock was devised by Richard of Wallingford (1292–1336) between 1327 and 1330, but no historical descriptions or schemes about this earliest one exists [23]. Until now, the most famous example is the 7-sided brass astronomical clock, as shown in Fig. 2.13, designed and created by Giovanni de' Dondi, a professor in the University of Padua and a clock maker between 1348 and 1364 [18, 24, 29, 31–33]. Seven dials of the upper section of this astronomical clock, respectively, demonstrate the primum mobile, Venus, Mercury, Moon, Saturn, Jupiter, and Mars. Figure 2.14 shows all of the reconstructed dials of de' Dondi's astrarium. The mechanism contained in each dial of the astronomical clock was an epicyclic gear train with pin-in-slot joints [34, 35]. This special design was common in the ancient astronomical mechanical devices and was essential for the anomaly motion.

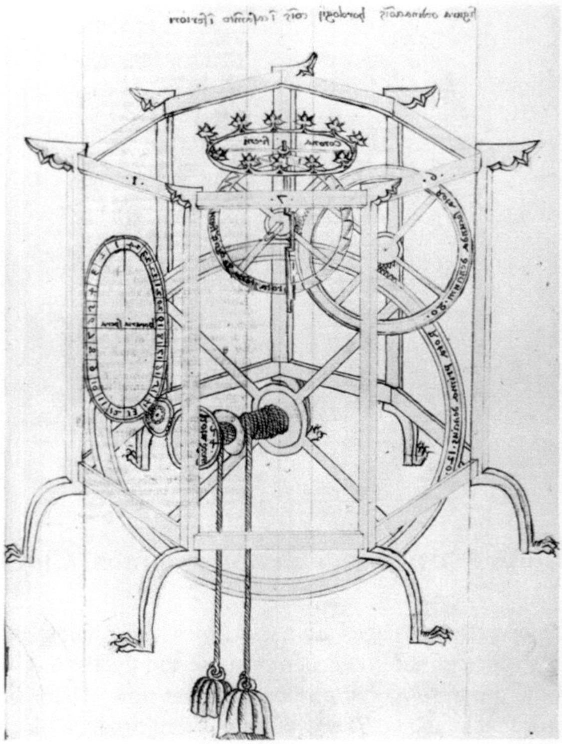

Fig. 2.13 Manuscript of astrarium by Giovanni de' Dondi [43]

In the fifteenth century, an unclear schematic drawing appeared in the manuscript of Leonardo da Vinci (1452–1519), as shown in Fig. 2.15a [29, 36]. The gear mechanism seemed to be a part of an astrarium and was supposed to serve the purpose of demonstrating the cyclic motion of Mars. Figure 2.15b shows the corresponding schematic drawing. There are 8 gears in this mechanism. Three of them (members 2, 6, and 9) are regarded as the carrier, and three of them (members 3, 7, and 10) are regarded as the input members for providing the input power. Therefore, this mechanism required three input sources to generate a controllable and constrained motion. By the viewpoint of modern mechanism, this mechanical device was a 9-bar mechanism with 14 joints and 2 degrees of freedom. The professional knowledge of mechanism is represented in Chap. 3. The chief concept of design should be satisfied with the Ptolemaic epicycle system. Certainly, it is obvious that the design of gear mechanism with pin-in-slot joints was applied again.

Besides, the demonstration function of planetarium was used for combining with other astronomical mechanical devices. For example, a French Gothic astrolabe, dated about 1300 and preserved in the Science Museum of London, has a similar purpose for displaying the relative positions of the Sun and the Moon in

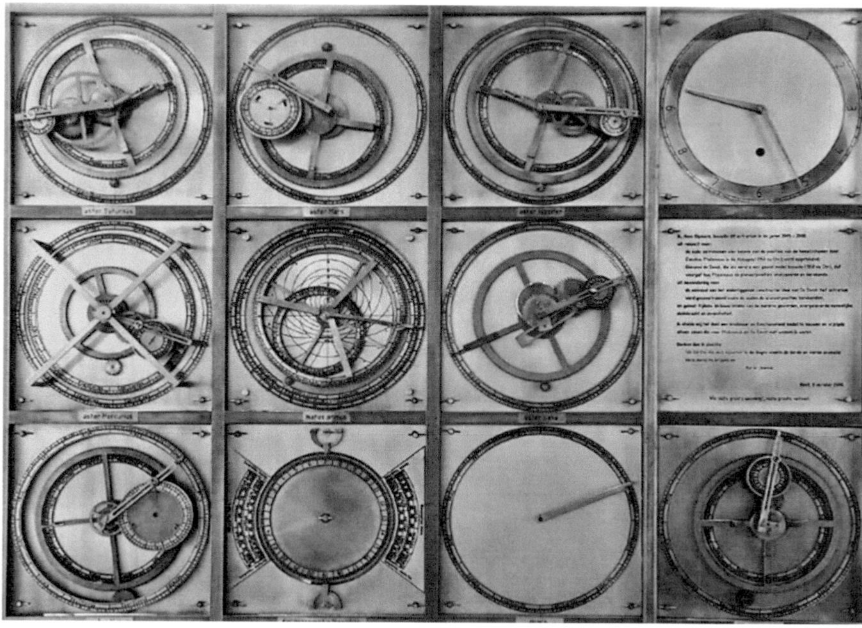

Fig. 2.14 Reconstructed dials of de' Dondi's astrarium [35]

the zodiac by the corresponding pointers driven by an arrangement of gears, as shown in Fig. 2.16 [24]. The gear mechanism was built over the front plate of the astrolabe. The existing situation of the Gothic astrolabe is quite complete, but lacking the Moon pointer. Therefore, its operation and functions are understood clearly.

In fact, the names "planetarium" and "astrarium" are frequently used to call the same device. In the manuscripts and treatises by de' Dondi, these two names appeared successively several times to mean his own astronomical clock. Therefore, in the study of astronomical mechanical devices in early times, the roles of planetarium and astrarium shall be regarded as the same.

2.7 Orrery

An orrery is a modern astronomical mechanical device that is designed based on heliocentric model. It displays the relative positions and motions of the planets and the Moon in the solar system [37–39]. The design and function of the orrery is very similar to the ones of planetarium or astrarium. Nevertheless, the orrery only refers to a display device whose motions of celestial bodies are around the Sun, i.e., the center of device is the Sun. The orrery was invented by clock makers

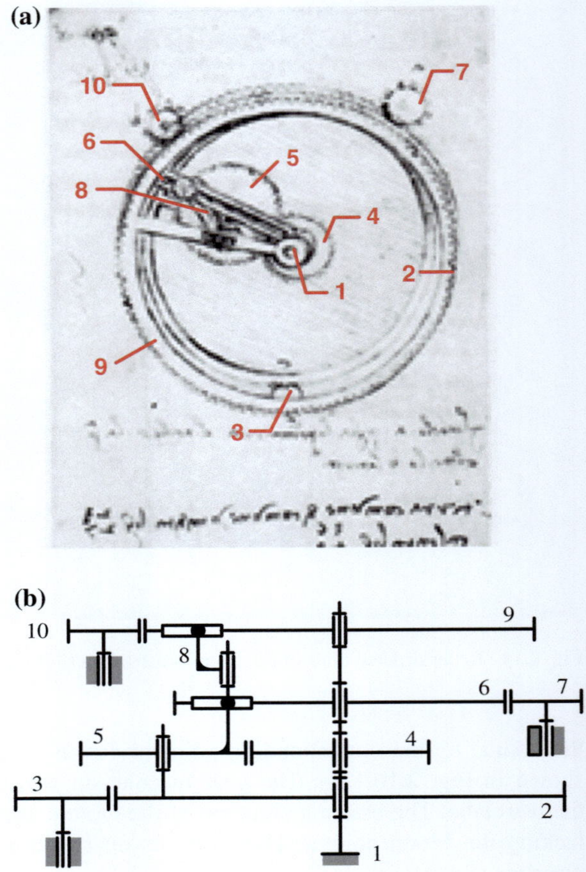

Fig. 2.15 The planetarium by Leonardo da Vinci. **a** Schematic drawing of historical manuscript [29, 36]. **b** Modern schematic drawing

George Graham (1673–1751) and Thomas Tompion (1639–1731) before 1719 and then gradually developed to be one of the achievements in the eighteenth century [37]. Such a device is typically driven by a gear mechanism, like a clock, and has a globe representing the Sun at the center as well as a planet at the end of each arm. The solar system is clearly represented in three-dimensional pointer system. The most important is that the revolution and the rotation of each planet can be understood easily.

2.8 Comparisons of Astronomical Instruments

A mechanical device is defined based on its functions and purposes. From the viewpoint of mechanical design, it is reasonable to believe that a common design for the specific purpose can be applied to various mechanical devices. A thorough

Fig. 2.16 French Gothic astrolabe, reprinted from Ref. [24], copyright 2015, with permission from Science Museum of London

investigation of the ancient astronomical mechanical apparatus is indispensable for understanding the historical development of mechanism designs for some specific output functions, such as computing, demonstration, and measuring.

According to available ancient astronomical mechanical devices (the Islamic calendrical device, the Byzantine sundial with calendrical gearing, and the Persian astrolabe with calendrical gearing), ancient mechanical design for the application of computing was ordinary gear trains. The excavation of some devices reveals that the design of gear mechanisms was workable and precise, even though the shapes of the teeth could be slightly different. In order to satisfy the calculations of different calendars, these astronomical apparatus with calendrical calculation had different teeth and numbers of gears to construct their gear trains.

The variation of moon phase is a cyclic astronomical phenomenon. It seemed to be regarded as one of the calendars. The function for displaying the variation of moon phase was usually attached to the calendrical device. The manners for showing the moon phase were particular and diverse. They depended on the relative motion of the Moon disk and the fixed plate for showing the moon phase. They could have a rotatable moon disk with full-black circles that were regarded as the moon shadow. Depending on the corresponding gearing, the moon disk was driven to gradually show the shadow in a hole of fixed case. Or, a Moon disk that was white and had a hole rotated over a fixed full-black plate for moon shadow. The fixed case still had a hole. And, the new moon appeared when the hole of Moon disk coincided with the hole of the fixed case. Furthermore, a Moon disk could be replaced by a Moon ball. This manner was rare, but existed in ancient astronomical apparatus.

For the function to demonstrate the motions of celestial bodies, the study of ancient astronomical apparatus concludes two different designs. One design is the ordinary gear train used in the Islamic astrolabe with calendrical gearing, the Byzantine sundial with calendrical gearing, the Persian astrolabe with calendrical gearing, the French Gothic astrolabe, and the orrery. Such a mechanical design represents the mean motions of heavenly bodies. Since these motions are almost anomaly, there are errors between the prediction positions of these apparatus and the actual observation positions in the night sky.

Another design is a planetary gear train with pin-in-slot joints. This design was applied in all dials of the Dondi's astronomical clock and the scheme of the manuscript by Leonardo da Vinci to generate the variable rate of output so as to satisfy the anomaly motions of the celestial bodies. The difference between the application examples is the numbers of the pin-in-slot joints. For these designs, the number of applied pin-in-slot joints is determined by their demonstrated planets and the universe model that the planets applied. According to the historical development of astronomy, both Dondi's astronomical clocks and the planetarium by Leonardo da Vinci were invented after the publication of the Ptolemaic system. Thus, it is confirmed that the demonstrations of planetary motions of these apparatus should adopt the Ptolemaic system. The numbers of pin-in-slot joints used made the designs satisfy the geometrical models of the astronomical theory at that time. Such a design concept was believed to have spread into Western scientific culture.

Except for the investigations of functions and mechanisms, the manufacture technique is also a point. For the ancient astronomical mechanical apparatus, gears were regarded as the soul of mechanisms. The difficulty of manufacturing was related to teeth shape, size, and teeth number of gears. The teeth shape of most early gears approximated to equilateral triangles, and the gear of Olbia is a particular example. Their shape and the required shape cutter were obviously simpler than modern gears. These early gears could be manufactured much easily, but the triangular shape made impact and vibration stronger during the engagement process of gears. In general, the size of the gears used in astronomical apparatus was so fine that they could be installed in a constrained case. As to the choosing of teeth number, it was absolutely related to the precision of scale division of a circle. Thus, all teeth could averagely surround the gear without deforming the shape.

Furthermore, coaxial arbors of gears were much popular in the ancient astronomical mechanical apparatus and clockwork. This design must consist of a hollow outer arbor, and the inner arbor could pass through it. The outer arbor and the inner arbor rotated, respectively. The manner for mounting a gear and an arbor was not constrained. Sometimes, some of the gears had squared holes by which they were fitted to their arbors. The relevant traditional technique was persisted generation by generation. At least, the design is used widely in the later content of this book.

2.9 Remarks

The research of ancient astronomical mechanical apparatus contributes to the understanding of the historical developments of their applications, mechanical designs, and relevant manufacturing techniques. It is essential that such an investigation provides the fundamental bases for constructing an astronomical mechanical apparatus with some specific functions.

Ancient astronomical mechanical apparatus, based on their functions, were classified into applications for observation, measuring time, measuring distance, computing, and demonstration. Except for separate ancient astronomical apparatuses, some of them are compound apparatus with multifunctions. For calendrical calculation and demonstration of celestial bodies' motions, several design examples are introduced and the relevant types of mechanisms and manufacturing technologies are analyzed.

In summary, it is believed that the design concepts to operate the motions with variable rates and the calendrical calculation were spread out in the Western world. And, the manufacture technique for these clock mechanisms had been preserved as a tradition.

References

1. Clagett M (1995) Calendars, clocks, and astronomy, ancient egyptian science, 2nd edn. American Philosophical Society, Philadelphia
2. Rossi C et al (2009) Chapter 2: measuring distance. In: Ceccarelli M (eds) Ancient engineers' inventions: precursors of the present. History of Mechanism and Machine Science, vol 8. Springer, The Netherlands, pp 13–25
3. Rossi C et al (2009) Chapter 3: measuring time. In: Ceccarelli M (eds) Ancient engineers' inventions: Precursors of the present. History of mechanism and machine science, vol 8. Springer, The Netherlands, pp 27–39
4. Rossi C et al (2009) Chapter 4: ancient computation devices. In: Ceccarelli M (eds) Ancient engineers' inventions: precursors of the present. History of mechanism and machine science, vol 8. Springer, The Netherlands, pp 41–59
5. Forbes G (1909) History of astronomy. G. P. Putnam, New York. (Republication) (2013) Forgotten Books. Available in the website. http://www.forgottenbooks.org. Accessed July 2014
6. Burgel BH (1911) Astronomy for all. Cassell and Company, Ltd., New York. (Republication) (2013) Forgotten Books. Available in the website. http://www.forgottenbooks.org. Accessed July 2014
7. Ducoudray G (1889) The history of ancient civilization. D. Appleton and Company, New York. (Republication) (2014) Forgotten Books. Available in the website. http://www.forgottenbooks.org. Accessed July 2014
8. Krehbiel DG (1990) Antique surveying equipment—Jacob Staff. In: The Ontario Land Surveyor. POB Publishing Company, Canton, Michigan. Also available in the website. http://www.surveyhistory.org/jacob's_staff1.htm. Accessed July 2014
9. Stern DP (2010) Spaceflight and the Earth's Magnetism, the Earth's motion in space: 5b. The Cross Staff. In: Physics. Astronomy of Educational Web Sites on Astronomy. Available in the website. http://www.phy6.org/stargaze/Scrostaf.htm. Accessed July 2014

10. Database of Water Clock of Ancient Egyptian (2014) Science Museum: Early Egyptian water clock, 1415–1380 BC (image number: 10326214). Available in the website. http://www.sciencemuseum.org.uk/images/i012/10326214.aspx. Accessed July 2014
11. Database of Cross Staff (2014) Celestial Navigation. Available in the website. http://www.math.nus.edu.sg/aslaksen/gem-projects/hm/0203-1-10-instruments/cross_staff.htm. Accessed July 2014
12. Cline DA (1999) Navigation: the cross staff. The Pilgrims & Plymouth Colony: 1620. Available in the website. http://www.rootsweb.ancestry.com/~mosmd/crstaff.htm. Accessed July 2014
13. Database of Jacob Staff (2014) National Technical Museum Prague. Available in the website. http://www.ntm.cz/en/en-exponat/jakubova-hul. Accessed July 2014
14. Museum of the History of Science (2006) The Astrolabe: An Online Resource. Available in the website. http://www.mhs.ox.ac.uk/astrolabe/. Accessed July 2014
15. Morrison JE (2007) The astrolabe. Janus, USA. Also available in the website (2010): The Astrolabe: An Instrument with a past and a future. http://www.astrolabes.org/. Accessed July 2014
16. Webster R (2007) Western astrolabe. In: Historic scientific instruments of the Adler planetarium series, vol 1. Adler Planetarium & Astronomy Museum, Chicago
17. Hayton D (2012) An introduction to the astrolabe. iBook Authors (online)
18. Mitchell TJ (2011) The astrolabe in theory and practice. In: Handout of the class. Available in the website. http://astrolabeproject.com/downloads/Astrolabe_in_Theory_and_Practice_Version_4.pdf. Accessed July 2014
19. Database of Ancient Machines (2009) Museo Galileo. Available in the website. http://catalogue.museogalileo.it/index.html. Accessed July 2014
20. Mintz D (2007) Timekeeping in the ancient world: sundials. Available in the website. http://www-history.mcs.st-and.ac.uk/HistTopics/Sundials.html. Accessed July 2014
21. Hays J (2012) Science and mathematics in ancient Egypt. In: Facts and Details. Available in the website. http://factsanddetails.com/world/cat56/sub365/item1923.html. Accessed July 2014
22. Hill DR (1985) Al-Bīrūnī's mechanical calendar. Ann Sci 42:139–163
23. Field JV, Wright MT (1985) Gears from the Byzantines: a portable sundial with calendrical gearing. Ann Sci 42:87–138
24. Bedini SA, Maddison FR (1966) Mechanical Universe: The Astrarium of Giovanni de Dondi. Trans Am Philos Soc 56(5):1–69
25. de Price DS (1959) On the origin of clockwork, perpetual motion device and the compass. US Natl Museum Bull 218(6):81–112
26. MHS Oxford (2013) Geared Astrolabe Animation (video). In: SHELF 3D. Available in the website. http://shelf3d.com/OiUgBeb2vtk#GearedAstrolabeAnimation. Accessed July 2014
27. Wright MT (2000) Greek and roman portable sundials: an ancient essay in approximation. Archive for history of exact sciences, Springer, Berlin, pp 177–187
28. Maddison F (1985) Early mathematical wheelwork: byzantine calendrical gearing. Nat News Views 314(28):316–317
29. Reti L (1974) The unknown Leonardo. McGraw-Hill, England
30. Knowledge of Wright's Reconstruction Design (20009) A byzantine sundial-calendar. Available in the website. http://hist.science.online.fr/antikythera/DOCS/FLORENCE2009/byzantine-sundial.htm. Accessed July 2014
31. de Dondi G (1397) The planetarium of Giovanni de Dondi: Citizen of Padua. (Republication) (1974) Antiquarian Horological Society, UK
32. Turner AJ (1975) The Tragical history of Giovanni de Dondi. J History Astron Sci Hist Publ Ltd 6:126–131
33. del Barrios MC (2011) A medieval planetarium as daily planner. In: Center for Medieval and Early Modern Studies. Available in the website. http://cmems.stanford.edu/node/833. Accessed July 2014

34. Croce CG (2012) Giovanni de Dondi ASTRARIUM. In: The home of antique clocks. Available in the website. http://www.clockmaker.it/inglepresentastrario.htm. Accessed July 2014
35. Database of Astrarium (2014) In: Watch-Wiki. Available in the website search. http://watch-wiki.org/index.php?title=Hauptseite/en. Accessed July 2014
36. Taddei M (2007) Leonardo da Vinci's Robots. Leonardo3, Italy. All relevant information available in the website: Leonardo3. https://www.leonardo3.net/leonardo/home_eng.htm. Accessed July 2014
37. Carlisle R (2004) Scientific American: inventions and discoveries. Wiley, New Jersey
38. Harris J (1757) The Description and UFE of the globes and the Orrery. B. Cole, and E. Cushee, London. (Republication) (2013) Forgotten Books. Available in the website. www.forgottenbooks.org. Accessed July 2014
39. Database of Orrery (2014) Smithsonian: National Museum of American History—Kenneth E. Behring Center. Available in the website. http://americanhistory.si.edu/collections/search/object/nmah_1185669. Accessed July 2014
40. Sabanski C (2014) The sundial primer. Available in the website. http://www.mysundial.ca/tsp/gnomon_height.html. Accessed July 2014
41. Database of Sundial Illustration (2013) Gnomon or Style. Available in the website. http://www.sundials.co.uk/tbgno.htm. Accessed July 2014
42. Database of Equatorial Sundials (2012) Equatorial Sundial, Sanford Center for Sky Awareness Observing Solar Cycles Using Sundials. Available in the website. http://www.wsanford.com/~wsanford/exo/sundials/equatorial_sundials.html. Accessed July 2014
43. Database of Horizontal Sundial (2007) Wikimedia Commons. Available in the website. http://commons.wikimedia.org/wiki/Main_Page. Accessed July 2014
44. Database of Non-planar Sundial (2014) Corbis images. Available in the website. http://www.corbisimages.com/stock-photo/rights-managed/IH049124/two-ancient-egyptian-sundials. Accessed July 2014
45. Database of Non-planar Sundial (2014) Ancient history encyclopedia. Available in the website. http://www.ancient.eu/Timekeeping/. Accessed July 2014

Chapter 3
Amazing Discovery of Archaeology

In the year of 1900, an unscheduled diving work accidentally resulted in a startling discovery of archaeological research, the oldest astronomical calculator—Antikythera mechanism. This device survived in the ancient Greece and dates about 100–150 BC. According to the excavation, it is understood that the device is composed of a great gear train with triangular teeth shape, and the sizes of most surviving gears are small. By the ingenious designs and the precise gears, the standards of manufacturing, science, and technology are believed to be highly developed in the era of Antikythera device. As a result, such an incident redefined the historical development of ancient Western machinery.

The beginning of this chapter describes this discovery story from the historical background of the incident. Then, the background related to this mechanical device is studied, along with the introduction of the known information of Antikythera device, including the exterior appearance, the device functions, and the corresponding interior mechanisms. In addition, the reconstruction designs by Yan and Lin have contributed to the describing of the interior mechanisms. The last part of this chapter presents the historical records by Cicero about the device.

3.1 Origination and Process of the Discovery

The discovery process of Antikythera device is a story full of surprise and luck. The sponge diving, an economical industry prevailing over the area in the nineteenth century, aroused the sophisticated mechanical invention form its deep sleep at 42 m beneath the surface of the Aegean Sea. In what follows, this marvelous discovery and the first discovery work are introduced. Except for the good luck of the first discoverer, some factors concluded by the historical background can possibly explain why the device was found after a long period of 2000 years.

J.-L. Lin and H.-S. Yan, *Decoding the Mechanisms of Antikythera Astronomical Device*, DOI 10.1007/978-3-662-48447-0_3

3.1.1 Historical Background of Salvage

Sponge diving has been an important and typical industry with a colorful historical background around the Mediterranean area and has enjoyed for a long history, since the waters around the islands with high temperature are very suitable for the sponges to grow. For several Greek islands of the Mediterranean in particular, the industry has been a fundamental revenue source for centuries. The diving for sponges brought social development to the islands [1–5]. The history of sponge diving dates back to antiquity. In some ancient writings by Aristotle, Plato, or Homerus, the sponge was mentioned as an object used for bathing; however, it is unknown exactly when this kind of usage of sponge began [5–7]. Now, the artificial sponge takes the place of the natural sponge in the daily life. The natural becomes the handiwork or the specialty of Greece, and the product can be easily found in a specialty shop in Greece, Fig. 3.1.

The sponge diving industry was more and more flourishing with the development of diving equipments and ships. At the beginning of sponge diving industry, the unique skill of sponge diving was the lifeblood of these Greek islands of the Mediterranean. The unique skill referred to the personal skill of "skin diving," which means that people dive the sponge without any equipment on [3–6]. Divers had no idea of the existence of diving equipments. Certainly, people at that time did not have the technique to manufacture the related equipments, such as helmets

Fig. 3.1 Sponge in the shop of Greece

or clothes for diving. The divers just dived into the depth of sea by the gravity of heavy stones. Therefore, the depth of diving was constrained by the limit of human bodies.

Until the tenth century, the applying and designing of diving equipments were gradually noticed, like the crest and the cloth [8]. The equipments made it possible for the sponge divers to work about 40–50 m in deep for a period of time. After 1865, the standard diving suit, "skafandro," enabled the spongers to gather larger sponges up to the depth of 70 m for longer times than was previously possible [4, 5]. Furthermore, forming a fleet of ships for sponge diving by a company enabled people to work at a farther distance for a long time. It resulted in the fact that sponge diving work with ship fleets almost ranged the entire Mediterranean [3].

Therefore, it is known that the development of diving equipments and the forming of sponge diving fleets affect the range of sponge diving, i.e., the depth and the distance. These are the critical factors in the finding of the Antikythera shipwreck, because people at that time were able to achieve the depth of shipwreck and work in the Mediterranean. Furthermore, it is logical to suggest that the factors may be parts of the reasons why the Antikythera shipwreck was not discovered until 1900.

3.1.2 Story of the Antikythera Finding

In 1900, captain Dimitrios Kontos and his crew were the sponge divers from the Symi island of Aegean. After their summer fishing work, the group of sponge divers was sailing home to Symi from Tunisia [9]. However, they could not go home smoothly owing to a violent storm at sea. These Greek sponge divers were driven off the course by gales and accidentally came to the island of Antikythera, midway between the Peloponnese and Crete, as shown in Fig. 3.2. This is a small island of about 20 km^2 in area.

After taking shelter on the island for a few days, they determined to work near the Antikythera Island. When the divers dived into the sea to search the sponge, they unexpectedly discovered an antique shipwreck at a depth of 42 m. In addition to a great quantity of bronze and marble statues, this shipwreck contained an unnoticed wooden box covered by limestone and moss [9–19]. Now, all excavations discovered in this shipwreck are preserved in the National Archaeological Museum of Athens [19].

In fact, this wooden box was not like other excavations that attracted many attentions. It stayed lonely in the corner of the museum until the second accidental finding in 1902. Valerios Stais, who was the director of the National Archaeological Museum of Athens at that time, noticed Antikythera device. As the shriveled excavations disrupted into the fragments and exposed the outer layers of bronze pieces, Stais discovered accidentally that an amazing arrangement of bronzed gears, triangles had been cut out of the middle of the wheel, which was hidden under the commonplace wooden box. According to the numbers of gears and the precision with which they were cut, he suggested immediately that

Fig. 3.2 Discovery journey of the Antikythera shipwreck [9]

the unknown excavation with complex mechanism might be a mechanical device for making accurate measurements or calculation. As far as all precious artifacts preserved in museums were considered, there was not anything like this design. Therefore, this strange device was just named "Antikythera mechanism" in accordance with the excavated position [9–19].

3.2 Introduction of the Excavations

The Antikythera shipwreck is a cultural treasury. Its relevant excavations are numerous, such as marbles, coins, vases, and weapons [9, 10, 19]. Antikythera device was a sophisticated mechanical device and seemed to be a miraculous one as a representative object of these excavations. All excavations are priceless and with cultural, art, or science background. Therefore, the ship full of valuable goods is suggested as a ship for paying tribute.

The excavation of Antikythera astronomical device, preserved in the National Archaeological Museum in Athens now, is a group of fragments, as shown in Fig. 3.3. The total number of the excavated fragments is 82 pieces, according to detection results [10–12, 19]. They include 7 major fragments (*A*–*G*) and 75 small fragments (no. 1–75). For the assembly positions of these excavated fragments, fragments *A*, *B*, *C,* and *D* belong to the top of the device, and fragments *E*, *F,* and *G* are below these 4 major fragments. The remaining fragments are uncertain, even they cannot be confirmed as parts of the device. The Antikythera device is

Fig. 3.3 Surviving fragments of the Antikythera astronomical device, reprinted from Ref. [12], copyright 2015, with permission from Nature

understood as a geared mechanism, and 29 gears has been confirmed from the surviving evidence. Most of the detected gears are contained in fragment *A*. A distinctive big gear that was hollowed as quarter divisions is visible in this fragment [10, 12, 20]. This is the most familiar representation of the Antikythera device. As to fragments *B*, *C*, and *E*, there exist several obvious segments of dials [10, 12, 19].

Size, area, thickness, and weight of each fragment were measured by modern equipments. The latest detection results indicated the largest fragment, fragment *A*, is approximately 224.209 cm^2, and the smallest one is approximately 0.146 cm^2. The measuring errors in area are estimated as no more than 0.01 cm^2. For the thickness of the metal portions of these fragments, they are almost between 1 and 2 mm. Some fragments are composed of several discernible layers. And the number of the contained layers is mostly about 6–7 layers. Besides, these fragments are quite light. The heaviest one, fragment *A*, is about 369.1 g. Basically, the primary geometric data was from microfocus X-ray computed tomography (CT), which is a completely nondestructive technique. These known-measuring data today are more accurate and completely depend on the ability of detectors and the identification of researchers [12].

The excavated fragments are the origin of all information of Antikythera device. Except for measuring their physical characteristics, it is detected that some of the excavated fragments belonged to different locations of the device, such as back door plate, front door plate, and lower back dial. They were full of glyphs and inscriptions. By the decoding process, these glyphs and inscriptions appear to be the label of graduations and the manual of the device, respectively.

In summary, the detection works of the excavation of Antikythera device are still under way. The further results will reflect on the decoding of the mechanisms inside this device in the future.

3.3 Known Antikythera Astronomical Device

Now, the Antikythera device has been gradually reconstructed with the updated detection results. The exterior appearance of the device that is now universally accepted in the academic community was built based on the reconstruction research by Freeth et al. [12, 17, 21].

Regarding the exterior appearance, this device is a bronze geared mechanism encased in a wooden box approximately 315 × 190 × 100 mm^3 and driven by hands through a crank attached on one side, as shown in Fig. 3.4. It has both front and back doors that are, respectively, covered with numerous inscriptions describing the information about the records of periodic astronomical phenomena and the calendars used commonly in Hellenistic Greek. Indeed, by the inscriptions with peculiar signs were decoded gradually, they were regarded as an instruction manual of the device [10, 12, 15–23].

3.3.1 Front Plate

Under the front door, the front plate has one metal dial which includes two concentric annuli with different scales, as shown in Fig. 3.5. The diameter of the dial is only a little less than the width of the case (the inner radius is about 62.5 mm; the middle radius is about 70 mm, and the outer radius is about 77.2 mm) [11].

(a) (b)

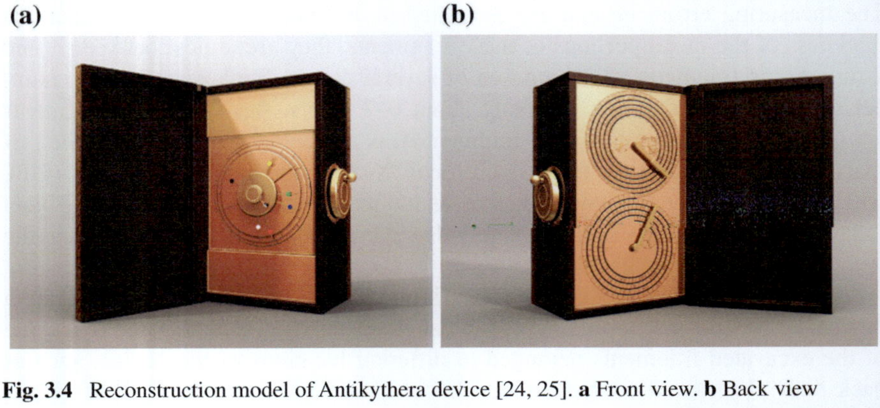

Fig. 3.4 Reconstruction model of Antikythera device [24, 25]. **a** Front view. **b** Back view

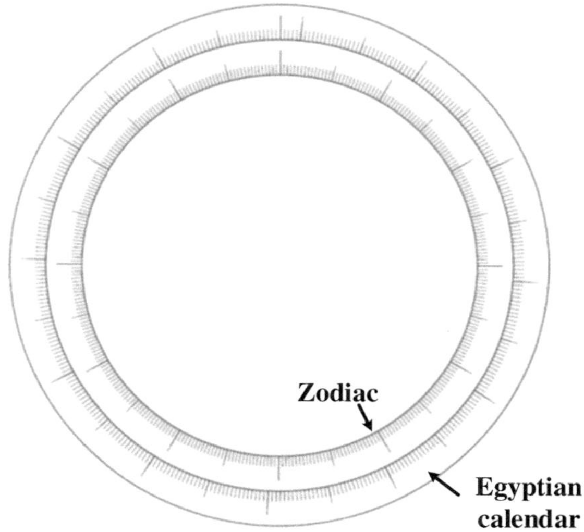

Fig. 3.5 Front plate of the Antikythera device

The inner scale is fixed and divided into 360 divisions in 30 groups. The groups are marked with names of the zodiac signs in Greece. The outer scale is movable and divided into 365 parts corresponding to the Egyptian calendar which has 12 months with 30 days each and a group of 5 extra days at the end of the year. A comparison with the true length of the year requires correcting the effect of the extra quarter day by turning the scale forward 1 day every 4 years. Besides, there are some inscriptions decoded as parapegma on the front plate. Parapegma is to describe the correspondence between the rising and the setting of particular stars and the civil date [12, 21]. Also, these inscriptions can be regarded as the manual to describe the operation of this device.

For this front dial, the front plate has several pointers that, respectively, display the daily motions of the date, the Sun, the Moon, and the five planets known at that time, including Venus, Mercury, Jupiter, Saturn, and Mars. Once the date is appointed, the positions of all designated celestial bodies and the corresponding moon phase can be revealed. Moreover, there is a black-and-white ball to rotate inside the hole of cover over the front dial. Such a special pointer is to display the changing of the moon phase during one month with the motion of the Moon. This function also appeared in other ancient astronomical instruments. These pointers are introduced in detail in Sect. 3.3.3.

3.3.2 Back Plate

The back plate has two parts, as shown in Fig. 3.6. The upper part, including one large dial and two subsidiary dials, is designed to express the calculations between different calendar systems and the records of celebrations in ancient Greece. The

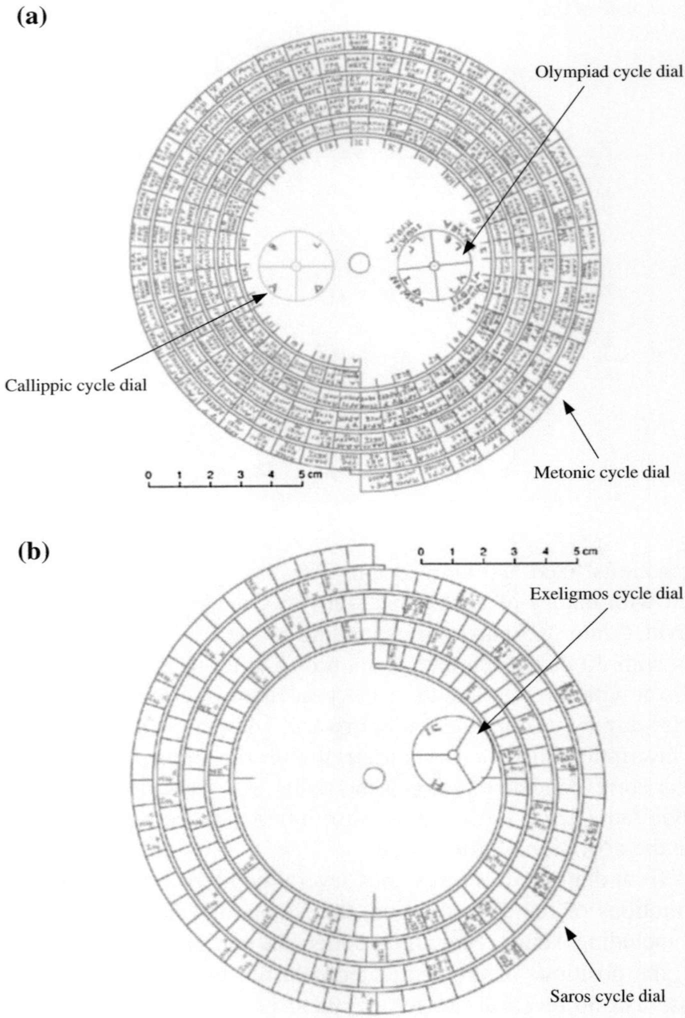

Fig. 3.6 Back plate of the Antikythera device, reprinted from Ref. [21], copyright 2015, with permission from Nature. **a** Upper back dial. **b** Lower back dial

large dial has 5 spiral turns, each of which has 47 divisions, to show the 235 synodic months of 19-year Metonic cycle. This cycle indicates the common multiple, approximately 6940 days, of the tropical year and the synodic month. When the date pointer displaying on the front dial rotates 19 turns, the pointer on this spiral dial finishes the 5 whole turns and needs setting again. Furthermore, there are still two subsidiary dials, respectively, divided into quadrants. One shows a calendar with the cultural importance for the Greeks that records the events associated with religious festivals at important cults or sites. In addition, the 4-year cycle of the Olympiad is included [21]. It is also suggested that one more subsidiary dial follows the 76-year Callippic cycle.

The lower part of the back plate that serves to predict the occurrence time of the solar and the lunar eclipses contains one large spiral dial and one subsidiary dial. The large dial has 4 spiral turns, on which 223 divisions are established, showing the 223 months of 18-year Saros cycle. When the date pointer rotates 18 turns, the pointer on the Saros dial would be finished the 4 whole turns. The subsidiary dial, divided into three parts, shows the 54-year Exeligmos cycle, i.e., triple Saros cycle. Furthermore, a set of numbers 0, 8, and 16 are labeled, respectively, on these sectors of the Exeligmos dial to indicate the shift time of an eclipse. Since the Saros eclipse cycle is not an integer, about 21,950 + 1/3 days, hours should be added in the labeled numbers to exactly predict the eclipse time [21]. The combination of these two dials not only predicts whether there is a lunar or solar eclipse in a special month but also indicates what time of day that eclipse would occur. Furthermore, due to each quarter turn of the Saros dial is equal to the full moon cycle, the cycle of changes in diameter of the full moon, the apparent diameter of the Moon can be indicated by the angle of pointer within each quarter turn of the dial.

In summary, each plate of Antikythera device includes several dials with various types and the corresponding pointers. Every dial, except for the Olympiad dial, displays clockwise. Once the daily input is given, these pointers can reveal the accurate calculation results on the dials.

3.3.3 Display Pointers

To observe the decoded pointers and dials of the Antikythera device, there are three types of pointers based on the styles and the motion directions, as shown in Fig. 3.7. In order to clearly introduce these pointer types, the mechanism has to be orientated. The orientation vertical to the dials of the front plate (or the dials of the back plate) is the axial direction, and the one parallel to the dials of the front plate (or the dials of the back plate) is the radial direction. Here, these three types of pointers of the Antikythera device are analyzed as follows.

3.3.3.1 Axial Rotation

Based on the orientation of mechanism, the pointers with axial rotation contain the most pointers of the Antikythera device. These pointers include the date pointer, the Sun pointer, the Moon pointer, the Mercury pointer, the Venus pointer, the Mars pointer, the Jupiter pointer, and the Saturn pointer in the front dials; the Olympiad pointer and the Callippic pointer in the back upper subsidiary dials; and the Exeligmos cycle pointer in the back lower subsidiary dials. These pointers are illustrated in Fig. 3.8a.

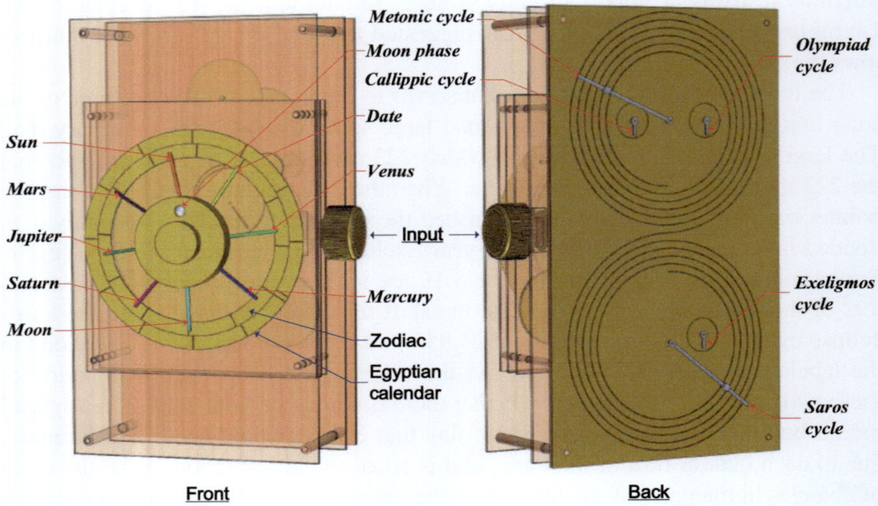

Fig. 3.7 Display pointers of Antikythera device [30]

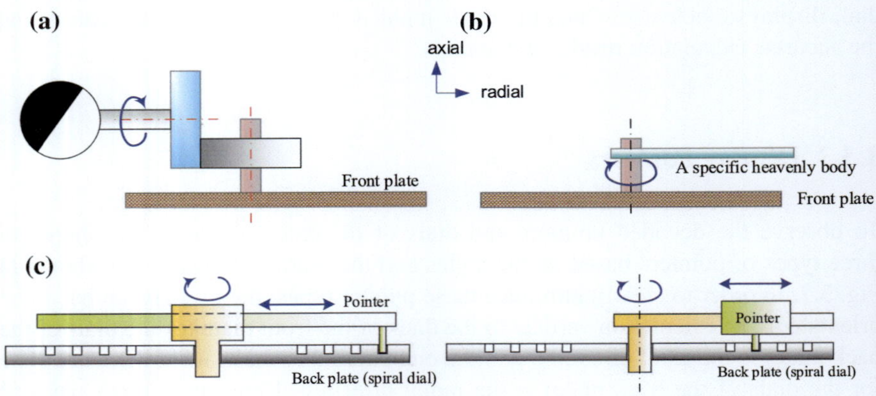

Fig. 3.8 Illustration of the pointer types [30]. **a** Axial rotation. **b** Radial rotation. **c** Axial rotation plus radial sliding

3.3.3.2 Radial Rotation

Only the pointer of the moon phase display device (the black-and-white ball) belongs to the pointer type with radial rotation, as shown in Fig. 3.8b. By the engagement of two gears whose rotational directions are orthogonal over the front, the motion of the Moon pointer can be transmitted to the black-and-white ball to display the moon phase corresponding to the Moon's position in the sky. The black-and-white ball should rotate around its shaft, i.e., to rotate along the radial direction based on the orientation of mechanism. Therefore, in most reconstruction models, it is shown that the moon phase display device was combined with a bronze disk, a bronze cover, and a gear train.

3.3.3.3 Axial Rotation and Radial Sliding

The pointer type with axial rotation and radial sliding, which contains the Metonic cycle pointer and the Saros cycle pointer, is the combination of a link and a slider. Two designs of this pointer type are presented by Freeth et al. [12, 17], as shown in Fig. 3.8c. The indicator of the design can be the slider or the link. Each design type could generate the spiral output motion along the spiral tracks. Also, these designs appeared in the research by Freeth et al. one after another.

The pointer type with axial rotation and the pointer type with axial rotation and radial sliding are absolutely confirmed. The corresponding pointers are clearly identified and reconstructed in the research by Freeth et al. There are slight differences between the appearance designs of pointers of the existing reconstruction models. However, some questions about the moon phase display device still exist. It is not confirmed that if the bronze disk is a pointer of the Antikythera device. This condition affects if the moon phase display device rotates around the axial direction of mechanism except for its own radial rotation. This work resolves the problem regarding how to restore all feasible designs of the moon phase display device.

3.3.4 Interior Structure of Mechanisms

As mentioned above, the exterior appearance and functions of Antikythera device are known through detecting and decoding the inscriptions on the surviving fragments. Corresponding to these exterior characteristics of the device, the interior mechanism of the device is complicated gear trains. There are several existing reconstruction designs of the interior structure in the historical development of the reconstruction research. The relevant story about reconstructing the interior mechanism for more than 110 years is introduced in Chap. 4. Here, a reconstruction model by Yan and Lin is used to illustrate the overall concept of the mechanism [24, 25], as shown in Fig. 3.9. This reconstruction design is generated by following the reconstruction research about the exterior characteristics and functions of the mechanism by Freeth et al. [10, 12, 17, 21] and a systematic reconstruction design methodology provided by Yan and Lin [24, 25]. The most important is that this design has the complete interior structure of mechanism to demonstrate all designated functions.

The interior mechanism of this reconstruction design is divided into six subsystems severing the exterior functions. These six subsystems are, respectively, referred to as the date subsystem, the calendrical subsystem, the eclipse prediction

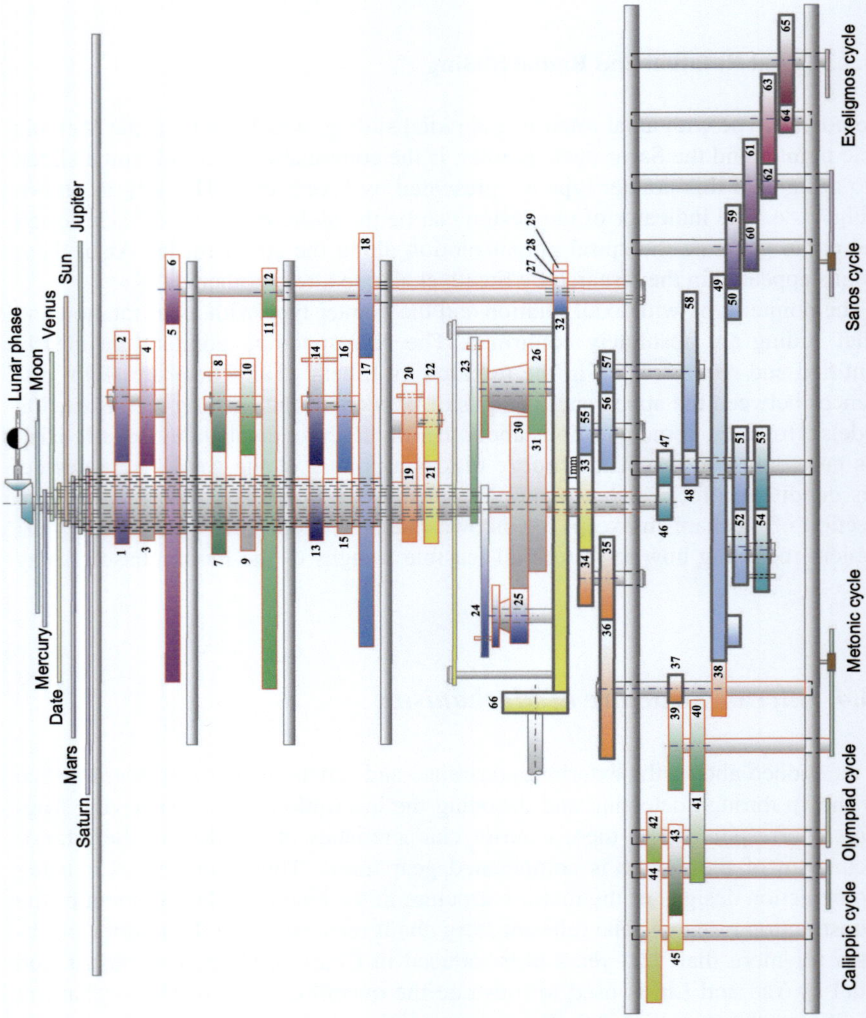

Fig. 3.9 A reconstruction design of the complete interior mechanism by Yan and Lin [24, 25]

subsystem, the lunar subsystem, the solar subsystem, and the planetary subsystem. The date subsystem is for the Egyptian calendar of the front plate. The eclipse prediction subsystem is for the Saros cycle and the Exeligmos cycle for the lower back dials. The calendrical subsystem is for the Metonic cycle, the Callippic cycle, and the Olympiad cycle of the upper back dials. The lunar subsystem, the solar subsystem, and the planetary subsystems are, respectively, for demonstrating the motions of the Moon, the Sun, and the five planets on the front zodiac dial. In addition, the completely lost mechanisms of the excavation and the other existing designs are composed of the solar subsystem and the planetary subsystem.

The known information of all mechanical elements is instructed clearly in accordance with the existing evidence and research. In the design shown in Fig. 3.9, the mechanical elements, which are confirmed from the surviving evidence, are in thick lines. And, the mechanical elements, which are estimated by the reconstruction works, are in thin lines. Thus, this complete reconstruction design indefinitely illustrates that some subsystems are already confirmed; some subsystems still have doubts and arguments; and some subsystems are completely lack of sufficient evidence for their designs. The detail mechanism analysis of this design and the design methodology to obtain this design are introduced in Chap. 5.

3.4 Relative Historical Background and Records

The purpose of studying historical records is to identify the origin of this ancient astronomical device and to provide the relevant information for the reconstruction design. Based on the understanding of Antikythera device, it is dated around 150–100 BC and has several functions of the astronomical calculations. It is confident that the device was not made by Archimedes, but might come from Syracuse in Sicily, the Corinthian colony where Archimedes had devised a planetarium in the third-century BC. Furthermore, it is speculative that the craftsmanship for making Antikythera device might be a heritage of manufacturing technique that originated with Archimedes in Syracuse. However, this attractive idea is waiting for proving [16, 21].

Through the investigation of historical literatures, there are almost no direct records that clearly point out the existence of Antikythera device in the extant literature dating to that period. The only relevant historical records merely appeared in two philosophical dialogs, "De republica" [26, 27] and "De Natura Deorum" [28, 29], by the greatest Roman orator Marcus Tullius Cicero. Some devices with the functions similar to those decoded form the surviving object of Antikythera device are described in the manuscripts of these two historical literatures.

In "De republica," Cicero mentioned that two devices predicting the movements of the Sun, the Moon, and the five planets known in that time were possessed by Roman general Marcus Claudius Marcellus as family heirlooms. They were built by Archimedes and considered as some types of ancient planetarium or orrery. The paragraph extracted from the manuscript of "De republica" is as follows:

> Philus: I can offer you, ... Listening one day to the recital of a similar prodigy, in the house of Marcellus, who had been his colleague in the consulship; he asked to see a celestial globe, which Marcellus's grandfather had saved after the capture of Syracuse, from this magnificent and opulent city, without bringing home any other memorial of so great a victory. I had often heard this celestial globe or sphere mentioned on account of the great fame of Archimedes. Its appearance, however, did not seem to me particularly striking. There is another, more elegant in form, and more generally known, moulded by the same Archimedes, and deposited by the same Marcellus, in the Temple of Virtue at Rome. But as soon as Gallus had began to explain, by his sublime science, the composition of this machine, I felt that the Sicilian geometrician must have possessed a genius superior to any thing we usually conceive to belong to our nature. Gallus assured us, that the solid and compact globe, was a very ancient invention, and that the first model of it had been presented by Thales of Miletus. That afterwards Eudoxus of Cnidus, a disciple of Plato, had traced on its surface the stars that appear in the sky, and that many years subsequent, borrowing from Eudoxus this beautiful design and representation, Aratus had illustrated them in his verses, not by any science of astronomy, but the ornament of poetic description. He added, that the figure of the sphere, which displayed the motions of the Sun and Moon, and the five planets, or wandering stars, could not be represented by the primitive solid globe. And that in this, the invention of Archimedes was admirable, because he had calculated how a single revolution should maintain unequal and diversified progressions in dissimilar motions. In fact, when Gallus moved this sphere or planetarium, we observed the Moon distanced the Sun as many degrees by a turn of the wheel in the machine, as she does in so many days in the heavens. From whence it resulted, that the progress of the Sun was marked as in the heavens, and that the Moon touched the point where she is obscured by the earth's shadow at the instant the Sun appears above the horizon. [26, 27]

In "De Natura Deorum," Cicero also mentioned that another such device was built by his friend Posidonius. Each one of the revolutions of which brings about the same movement in the Sun and the Moon and the five wandering planets as is brought about each day and night in the heavens.

> Why, if any one were to carry into Scythia or Britain the globe which our friend Posidonius has lately constructed, each one of the revolutions of which brings about the same movement in the Sun and Moon and five wandering stars as is brought about each day and night in the heavens, no one in those barbarous countries would doubt that that globe was the work of intelligence. Yet the Epicureans doubt as to whether the universe, from which all things arise and are created, was itself the result of chance or some kind of necessity, or of intelligence and the divine mind, and they think that Archimedes did more in imitating the revolutions of the sphere than nature did in producing them, although the original was wrought with far more cunning than the imitation. [28, 29]

Since the accounts in Cicero's writings lacked the drawings of these mechanical devices, they could not prove if these designs are Antikythera device. Furthermore, it is uncertain that these accounts were truth or fabrications. The authenticities of these devices also should be deliberated. If all accounts in Cicero's writings are true, it is surprising that the craftsmanship for manufacturing the Antikythera device with such a complex mechanism and precise calculation functions had been developed in Archimedes' times, the third-century BC, and still survived in Posidonius' times at least.

In summary, although very few historical records related to the Antikythera astronomical device were on considering, it is probable that this sophisticated invention is not a unique mechanical device.

3.5 Remarks

Except for the copious treasure, the discovery of the Antikythera shipwreck in 1900–1901 accidentally revealed the ingenious Antikythera device, which is regarded to have ever existed in the world of ancient Greece. The excavation is a group of fragments. Under the corroded appearance of the excavation, this sophisticated mechanical device composed of a great quality of gears with triangular teeth shapes and fine sizes indicating the superior standard of manufacturing technique and science in the subject's era. It is much amazing that the manufacturing level of the device is almost close to the manufacturing level of the astronomical clocks in the thirteen century, such as the clock in Padua by Giovanni de' Dondi between 1348 and 1364.

For investigating the origination of mechanisms, it is really unusual that there were no historical records of the Antikythera device. However, two philosophical dialogs by Cicero divulged the existence of such ancient astronomical apparatuses. These writings could be exaggerated or imaginative and were not the direct evidence of the Antikythera device. If they are reliable, such a craftsmanship and science for building astronomical apparatuses would be a tradition around the area.

The Antikythera device is an astronomical mechanical calculator. The front plate shows the cyclic motions of celestial bodies on the zodiac and its daily motion. The back plate shows the lunisolar calendars and the prediction of eclipses, respectively, on the upper and lower systems. According to Yan and Lin's reconstruction designs, its mechanical structure contained six subsystems to demonstrate the motions of the Sun, the Moon, and the five planets, to indicate the date, to display the moon phase, to calculate the calendars, and to predict the eclipses.

References

1. Warn F (2000) Bitter Sea, The real story of Greek Sponge Diving. Guardian Angel Press, South Woodham Ferrers, UK
2. Rützler K(1996) Sponge diving—Professional but not for profit. In: Lang MA, Baldwin CC (eds) Methods and techniques of underwater research. Proceedings of the American Academy of Underwater Sciences the 16th Scientific Diving Symposium, Washington DC, pp 183–204

3. Hendrikse S, Merks A (2009) Diving the Skafandro suit. Available in the website. http://www.divingheritage.com/greecekern2.htm. Accessed July 2014
4. Kalymnos Sponge Diving (2014) Greeka.com: The Greek island specialists. Available in the website. http://www.greeka.com/dodecanese/kalymnos/sponge-diving-tradition.htm. Accessed July 2014
5. The Diver training Network (2010) The Sponge Divers. In: The Sponge Divers of Kalymnos. Available in the website. http://www.kalymnosdivingfestival.com/index_files/Page1172.htm. Accessed July 2014
6. Few Things about the History of the Natural Greek Sea Sponges (2014) Greek Sponge Divers, Cambridgegreekcorner. Available in the website. http://www.cambridgegreekcorner.co.uk/. Accessed July 2014
7. Voultsiadou E (2007) Sponges: an historical survey of their knowledge in Greek antiquity. J Mar Biol Ass UK 87:1757–1763
8. Database of Diving Suit (2014) GTP headlines. Available in the website. http://news.gtp.gr/2010/08/01/kalymnos-an-ideal-destination-for-all/. Accessed July 2014
9. Marchant J (2008) Decoding the heavens—solving the Mystery of the World's first computer. William Heinemann, Great Britain
10. Databased of the Antikythera Mechanism (2014) The Antikythera Mechanism Research Project. Available in the website. http://www.antikythera-mechanism.gr. Accessed July 2014
11. de Price DS (1974) Gears from the Greeks: The Antikythera Mechanism—A calendar computer from ca. 80 BC. Science History Publications, New York
12. Freeth T et al (2006) Decoding the Ancient Greek astronomical calculator known as the Antikythera mechanism. Nature 444:587–891
13. Charette F (2006) High tech from ancient Greece. News Views Nat 444(30):551–552
14. Marchant J (2006) In search of lost time. News Feature Nat 444(30):634–638
15. Wright MT (2007) The Antikythera mechanism reconsidered. Interdisciplinary Sci Rev 32(1):27–43
16. Ball P (2008) Complex clock combines calendar. News Nat 454(31):561
17. Freeth T (2009) Decoding an ancient computer. Sci Am 301(6):79–83
18. Moussas X (2011) Chapter 8—The Antikythera mechanism: a mechanical cosmos and an eternal prototype for modelling and paradigm study. In: Kokkotas PV, Malamitsa KS, Rizaki AA (eds) Adopting historical science knowledge production to the classroom. Sense Publishers, Boston, pp 113–129
19. Kaltsas N, Vlachogianni E, Bouyia P (2012) The Antikythera Shipwreck: the ship, the treasures, the mechanism. Hellenic Ministry of Culture and Tourism, National Archaeological Museum, Athens
20. Wright MT (2004) The scholar, the mechanic and the Antikythera Mechanism. Bull Sci Instrument Soc 80:4–11
21. Freeth T et al (2008) Calendars with olympiad display and eclipse prediction on the Antikythera mechanism. Nature 454:614–617
22. Wright MT (2002) In the steps of the master mechanic. In: Ancient Greece and the Modern World. Proceedings of a conference in Ancient Olympia, 2002, University of Patras, Patras, Greece, pp 86–97
23. Wright MT (2012) The front dial of the Antikythera mechanism. In: Proceedings of the HMM2012. Springer, Netherlands, No. 15, pp 279–292. HMM 2012: International Symposium on History of Machines and Mechanisms, VU University Amsterdam, The Netherlands
24. Lin JL, Yan HS (2014) Historical development of reconstruction designs of Antikythera mechanism. Presented in proceedings of the 9th international conference on history of mechanical technology and mechanical design, Tainan, 23–25 Mar 2014. (Applied mechanics and materials, vol 163, 1–62012.04)
25. Lin JL (2011) Systematic reconstruction designs of Antikythera mechanism. Dissertation, National Cheng Kung University, Tainan, Taiwan

26. Cicero MT (54 BC) De Republic. English edition: M.T. Cicero (1999) On the Commonwealth and on the Laws (trans: J.G. Zetzel). Cambridge University Press, United Kingdom
27. Cicero MT (54 BC) De Republic. English edition: M.T. Cicero (2012) Commonwealth (trans: F. Barham). The Online Library of Liberty, A Project of Liberty Found, Inc. Available in the website
28. Cicero MT (45 BC) De Natura Deorum. English edition: M.T. Cicero (2012) On the Nature of the Gods (trans: F. Brooks). The Online Library of Liberty, A Project of Liberty Found, Inc. Available in the website. http://oll.libertyfund.org/index.php?Option=com_staticxt&staticfile=show.php%3Ftitle=539&layout=html Accessed July 2014
29. Cicero MT (45 BC) De Natura Deorum. English edition: M. T. Cicero (1933) De natura deorum, Academica (trans: H. Rackham). Harvard University Press, London
30. Lin JL, Yan HS (2014) Decoding the moon phase display device of the front dial of the Antikythera Mechanism. Paper also presented in Third IFToMM on mechanism and machine science, Tianjin, China, 9–10 July 2014

Chapter 4
Modern Reconstruction Research

All decoded information of the Antikythera device has been studied through hard investigations for a long time. Scholars in the time of the discovery failed to explain how this device can be operated. It was really puzzling trying to figure out what this device with the complicated mechanisms and the fine elements was. The fragmentary excavation, the sophisticated mechanical designs, and the immature technology make the device more mysterious. These mysterious characteristics have attracted scholars to decode and reconstruct such an ancient invention for more than 110 years.

The knowledge about Antikythera device is attributed to a number of individuals and research groups, who have focused on the restoring research, such as A. Rehm, D. Price, A.G. Bromley, M. Edmund, P. Morgan, M. Wright, T. Freeth et al., Lin and Yan, etc. In addition, more and more researchers invest their professional knowledge in decoding the device. In this chapter, historical developments of the reconstruction research of Antikythera device are introduced. Research results and contributions in different periods are presented. Various reconstruction designs are compared. And, their differences are identified clearly with the known design of Antikythera device.

4.1 Early Mentions

Not only Greek people but also researchers around the world successively studied the Antikythera astronomical device since Stais's finding from 1902 to 1920. The archeological journal dedicated a series of issues to the findings in the shipwreck. However, they were unable to present a coherent explanation due to the limitation of technology at that time. Most of their study results mentioned what kind of instruments the Antikythera device was. Either the exterior appearance or the interior mechanisms were never identified.

J.-L. Lin and H.-S. Yan, *Decoding the Mechanisms of Antikythera Astronomical Device*, DOI 10.1007/978-3-662-48447-0_4

Around 1905, A. Rehm, a German philologist, was the first scholar to define that the Antikythera device was an astronomical calculator. In accordance with his opinion, this device should be a planetarium resembling the sphere of Archimedes or Posidonius that Cicero described in 100 BC [1–3]. At the same time, the naval historian K. Rados sided with Rehm's viewpoint to claim that Antikythera device was too complex to be an astrolabe. However, there were alternative arguments against above concept of spherical planetarium. In 1910, P. Rediadis, a lieutenant of the Royal Navy, took a thorough description of the instrument, through the observation and sketch of the fragments. He claimed that the device was a kind of astrolabe in ancient Greece and was used to measure the altitude of the stars in the sky adjusted to the season. By any means, the embryo of Antikythera device was nothing during this time period.

4.2 Reconstruction Work by Price

Following decades of decoding work, the first reconstruction design with the concrete exterior appearance and systematic interior mechanisms was delivered by Derek John de Solla Price (1922–1983), as shown in Fig. 4.1. His research about the reconstruction of Antikythera device almost spanned the years 1959–1981. In addition, the book entitled "Gears From the Greeks: The Antikythera Mechanism—A Calendar Calculator" depicting all his contributions to the reconstruction work of the Antikythera device was published in 1974 [4]. The physical model corresponding to his reconstruction design was successfully built, as shown in Fig. 4.1a.

D. Price, a British science historian and prolific writer, focused on studying the ancient astronomy and the related ancient astronomical devices during his research life. The research materials contain ancient astrolabes, ancient sundials, Diovanni de Dondi's planetarium, Islamic celestial globes, bronze metallurgy, etc. In 1951, he undertook the investigation of the Antikythera device. Undoubtedly, all of his previous research subjects are completely contributed to his investigation of Antikythera device, especially to his master ancient astronomical instruments.

In the reconstruction work of Antikythera device, Price cooperated with G. Stamiris and C. Karakalos, using the X-ray tomography, an image technology in the modern time, to thoroughly detect the excavated fragments. G. Stamiris and C. Karakalos detected the possible gears and teeth from a group of tomography images. Moreover, Price evaluated detected results and assembled its possible structure. According to the understanding of inscriptions at that time, Price's hypothesis was that the device was constructed in the academy of Poseidonios from Rhodes, a core area of not only rhetoric, art, and astronomical studies but mechanics also. In addition, Price believed that the ship carrying the device to Rome was transporting other scientific inventions as well.

The exterior appearance of the Price's design is shown in Fig. 4.1a, and the front plate had a dial composed of two concentric annuli [4, 5]. The inner annulus

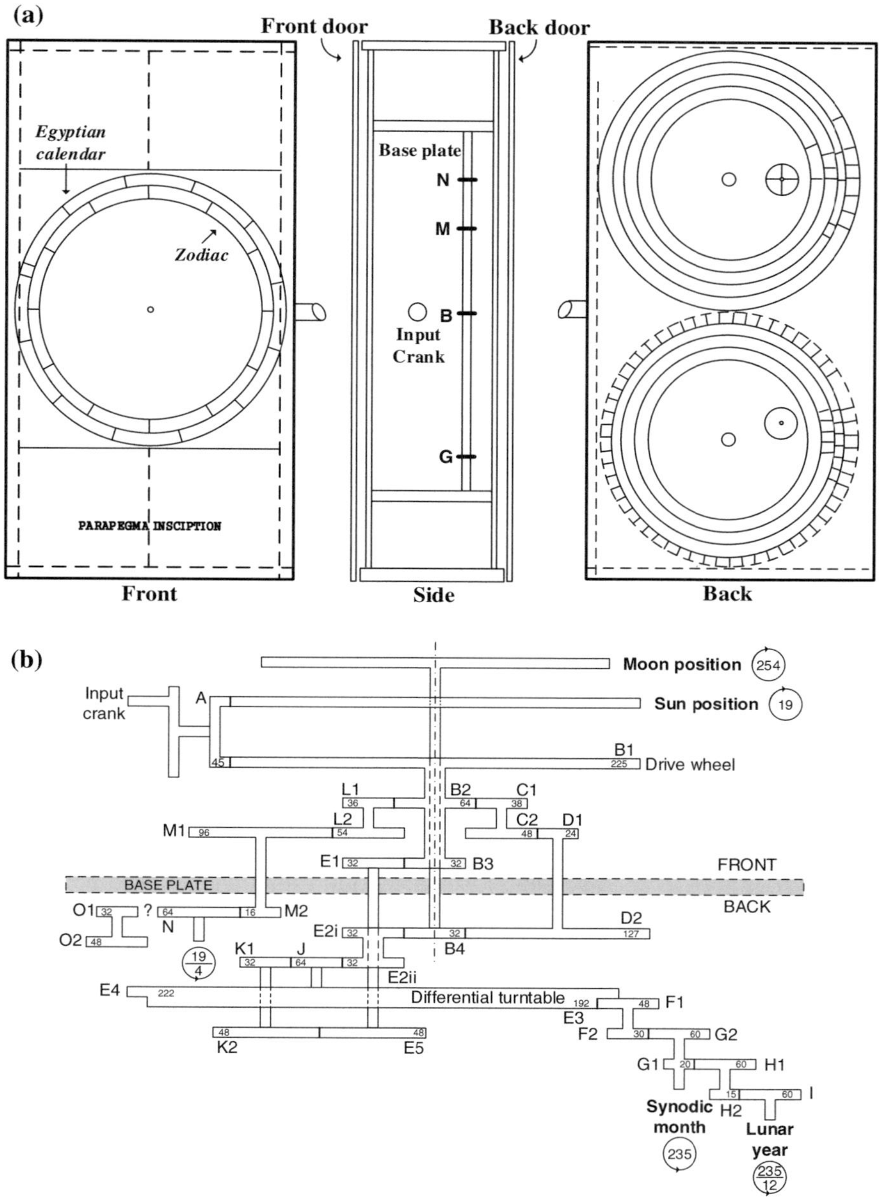

Fig. 4.1 Reconstruction design by Price. **a** Exterior reconstruction design and reconstructed physical model [4, 37]. **b** Interior reconstruction design [4]

was fixed as zodiac, and the outer one was rotatable as calendar. Both of them were graduated with short marks at degree intervals and longer marks every 30°. Therefore, the inner annulus had 12 major divisions for the well-known 12 zodiac constellations. The outer annulus had 12 major divisions for the 12 months of a year. The back plate included two concentric dials and two subsidiary dials. The upper back dial was supposed to have four concentric and rotatable annuli. The inscriptions related to the upper concentric dial were not identified, and its purpose was unknown. Price's scheme incorporates the suggestion that the pointer should turn once every 4 years. Alternatively, he suggested that it might have turned 5 times in the 19-year Metonic cycle, with synodic months indicated by the division of dial into 47. The lower back dial was supposed to have three concentric and rotatable annuli. Also the fixed outer limb of the lower dial was divided. Its purpose was inferred to display the synodic months. Besides, there were two subsidiary dials inside these two concentric dials of the back plate, respectively. The upper subsidiary dial was divided into 4 divisions, but its purpose was uncertain. The lower subsidiary dial inside the lower dial was supposed to show the lunar year.

As far as the exterior appearance and the functions are concerned, Price's reconstruction design of the Antikythera device is not the same as the known device mentioned in Chap. 3. These reconstructed results have the same design of the front dial. Nevertheless, their back dials are completely different, whatever the function definitions or the types of the dials.

Figure 4.1b shows the interior mechanism of the Price's reconstruction design [4, 5]. This restored mechanism corresponds to his identified purposes of the exterior dials. The reconstructed mechanism can roughly be regarded as three parts. One part is to operate the Sun pointer on the front dial. It contains a gear, which is the same as the main wheel, and a contrate gear connected to the crank handle. Another part is a differential gear train containing 6 links, 5 revolute joints, and 3 gear joints. This gear train is to operate the Moon pointer. The third part is an ordinary gear train containing 5 links, 4 revolute joints, and 3 gear joints to demonstrate the 235-synodic month and the lunar year on the lower back dials. Since the upper back dials are unknown in Price's design, the corresponding gear mechanism is incomplete. These gear mechanisms about the solar motion and the lunar motion are not constructed based on the geocentric theory mentioned in Chap. 1. In addition, these astronomical motions are demonstrated with the concept of mean rates, not true motions, i.e., anomaly motions. They are slightly regarded as the representation of the year and the month of the Egyptian calendar. Furthermore, the gear mechanism for the 235-synodic month and the lunar year is for calendrical calculations. It shows the 19-year Metonic cycle and the relationship between the solar year and the lunar year. Nevertheless, this function is merely speculated.

4.3 Reconstruction Work by Edmund and Morgan

Price's reconstruction design did not cover several lost and incomplete parts in the interior mechanisms of Antikythera device. Based on Price's design, Mike Edmunds and Philip Morgan presented a possible type of gear mechanism focusing on the lost and incomplete parts in 2000 [6]. Two gear mechanisms were added into the device to demonstrate the Venus' motion and the Mar's motion, respectively, as shown in Fig. 4.2. The gear mechanism for the Venus was located at the front lost part of Price's design (over the driving wheel), and it displays the Venus' motion on the front dial. The gear mechanism for the Mars was located at the back, an incomplete part of the Price's design. In other words, the gear mechanism of the calendrical subsystem was still nonexistent in Edmund and Morgan's design. It was misunderstood as the gear mechanism for the Mars' cyclic motion.

Except for the differences of the teeth, these two mechanisms are the same. Figure 4.3 shows the common design of these two added gear mechanisms. For such a design regarding the motions of Venus and Mars, its characteristics of topological structure are concluded as follows:

1. It is a planar 4-bar mechanism with 5 joints and 1 degree of freedom.
2. It has a ground link (member 1, K_F), an input link (member 2, K_I), an output link (member 3, K_O), and a transmission link (member 4, K_T).

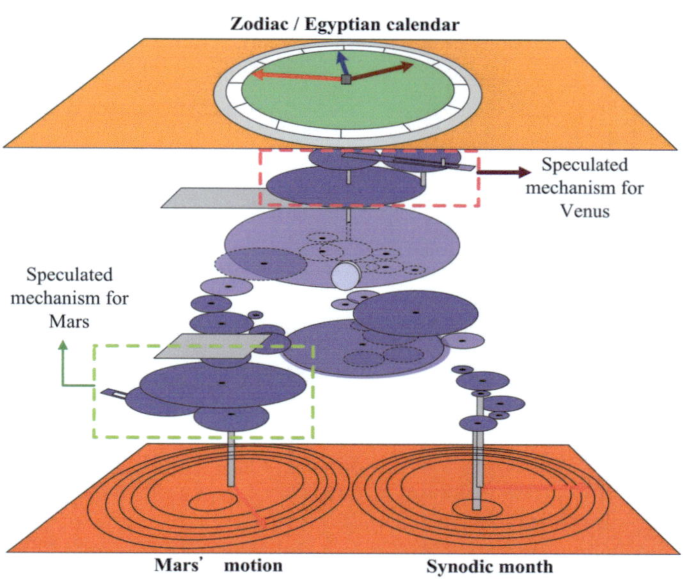

Fig. 4.2 Interior reconstruction design by Edmund and Morgan [6]

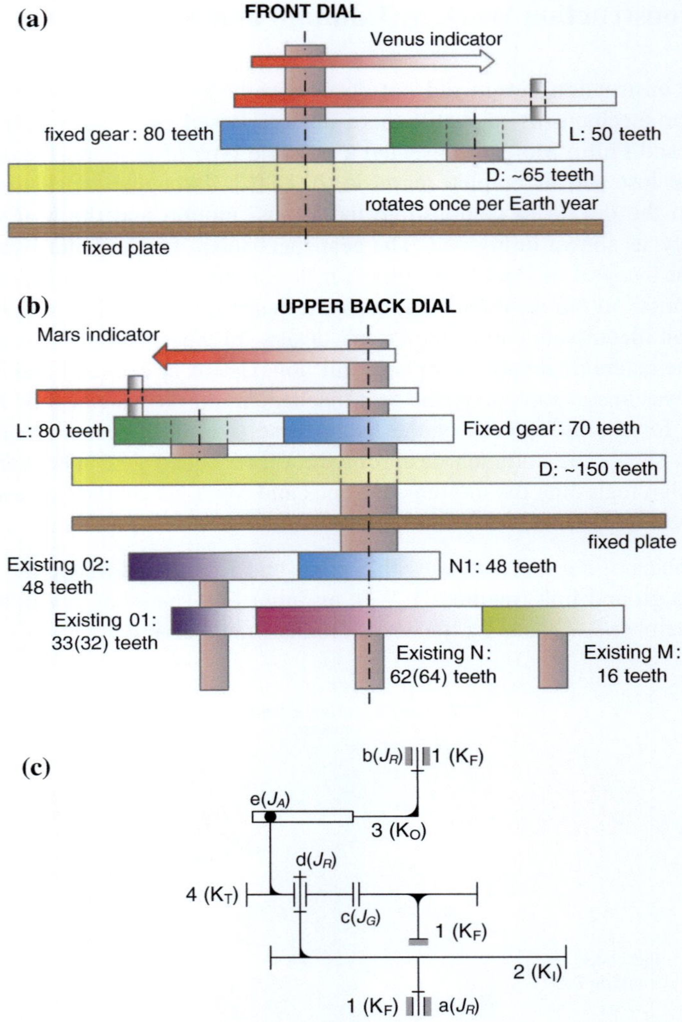

Fig. 4.3 Two added parts of the interior design by Edmund and Morgan. **a** Front partial mechanism [6]. **b** Back partial mechanism [6]. **c** Schematic drawing

3. It has 3 revolute joints (joints *a*, *b*, and *d*; J_R), 1 gear joint (joint *c*; J_G), and 1 pin-in-slot joint (joint *e*; J_A).
4. It is an epicyclic gear train with a pin-in-slot joint, which is able to generate the anomaly motion.
5. All gears of the epicyclic gear train are external gears exclusively.

The topological matrix, which is defined in Sect. 5.4.1, of the lost structure for Venus' motion (M_{T}) is as follows:

$$M_{\mathrm{T}} = \begin{bmatrix} K_{\mathrm{F}} & J_{\mathrm{R}} & J_{R} & J_{\mathrm{G}} \\ a & K_{\mathrm{I}} & 0 & J_{\mathrm{R}} \\ b & 0 & K_{\mathrm{O}} & J_{\mathrm{A}} \\ c & d & e & K_{\mathrm{T}} \end{bmatrix}$$

Furthermore, according to the above analysis of topological structure, the design of an epicyclic gear train with a pin-in-slot joint is used to reconstruct the motions of heavenly bodies. Such a design concept is very general in other ancient astronomical instruments with the demonstration function, such as the planetarium mentioned in Chap. 2. Moreover, based on the design concept of the added gear mechanisms, it is believed that the motions of Mars and Venus are identified as anomaly motions. They have to satisfy the epicyclic system by Apollonius in the Hellenistic age, i.e., the planetary theory introduced in Sect. 1.3.3.

4.4 Reconstruction Work by Wright

Michael T. Wright (1948–), an England historian of technology, has focused his research on studying the Byzantine instruments with the calculation function and the display of the moon phase in the past years [7]. He has abundant experiences about the reconstruction research of ancient mechanical devices, before undertaking the research of Antikythera device.

Wright has made a lot of efforts for restoring the Antikythera device ever since 2000 [8–21]. He did a study of the excavated fragments of Antikythera device with Allan George Bromley. They figured out that some of Price's arguments were unsound to follow based on his reconstruction design. By the development of the image-processing technology, they used linear tomography (advanced radiography) to redetect the original fragments, then amassed a great quantity of data. The decoding and analysis work of the image-processing data to reconstruct the mechanism was really a time-consuming and thorny investigation. Therefore, Wright was preoccupied to deliberate the possible designs for the missing parts of the mechanism by his finding of some original inscriptions [9, 10, 12–14].

Wright believes that Antikythera device could display the cyclic motions of planets known at that time on its front zodiac dial. The positions of specific planets in one day could be estimated by the mechanism. It was a mechanical design that integrated space condition with time condition. The mechanism should be a kind of ancient planetariums. Therefore, Wright built a physical model based on his previous experience on the reconstruction research of ancient astronomical instruments in 2005, before they found a resolution for these data [10–13, 21].

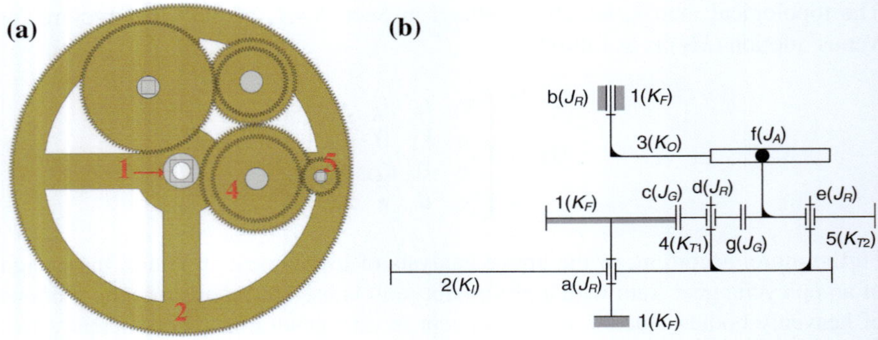

Fig. 4.4 Reconstructed partial mechanism for solar motion. **a** Physical model [13]. **b** Schematic drawing

Wright successfully rebuilt the physical model of the front dial of the Antikythera device with the display functions of planetariums [14]. The gear mechanisms for simulating the celestial bodies' motions were regarded as the lost parts of the interior design. It is emphasized that these gear mechanisms for the celestial bodies' motions were not detected in the surviving fragments. There was not any former arguments related to these lost parts. In addition, these gear mechanisms are epicyclic gear trains with a pin-in-slot joint and can generate the anomaly output motions. The reconstruction model of the front dial can display the motions of the Sun, the Moon, and the five planets. All of these motions of heavenly body are anomaly motions, and their design concepts should depend on the epicyclic system or the eccentric system developed in the era of Antikythera device mentioned in Sect. 1.3. Undoubtedly, as comparing to Price's reconstruction design, Wright's reconstruction model provides a possible design of the front part of the interior mechanism to demonstrate the planetary motions, except for the solar motion and the lunar motion. These functions were never presented in Price's reconstruction design.

Figure 4.4a is the reconstructed mechanism to generate the motion of the Sun, i.e., the solar subsystem, and Fig. 4.4b shows the corresponding kinematic drawing. Based on the concept of modern mechanisms, the characteristics of the topological structure of the reconstructed solar mechanism are as follows:

1. It is a planar 5-bar mechanism with 7 joints and 1 degree of freedom.
2. It has a ground link (member 1, K_F), an input link (member 2, K_I), an output link (member 3, K_O), and 2 transmission links (members 4 and 5, K_{T1} and K_{T2}).
3. It has 4 revolute joints (joints a, b, d, and e; J_R), 2 gear joints (joints c and g; J_G), and 1 pin-in-slot joint (joint f; J_A).
4. It is an epicyclic gear train formed exclusively by external gears.
5. The input link, as the carrier, rotates clockwise at the period of a tropical year.

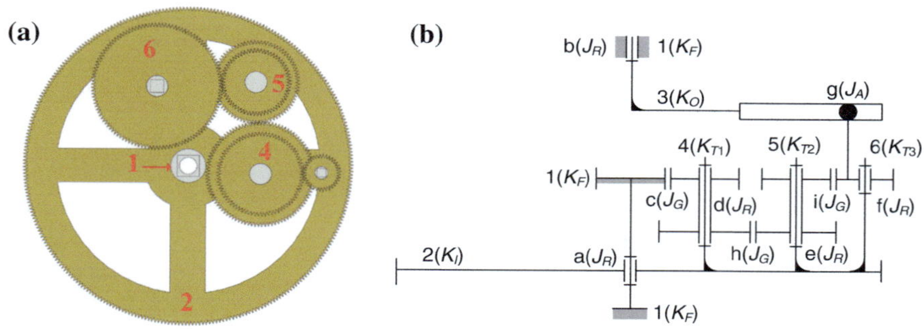

Fig. 4.5 Reconstructed partial mechanism for Venus' motion. **a** Physical model [13]. **b** Schematic drawing

And, the topological matrix of the solar subsystem (M_T) is as follows:

$$M_T = \begin{bmatrix} K_F & J_R & J_R & 0 & J_G \\ a & K_I & 0 & J_R & J_R \\ b & 0 & K_O & J_A & 0 \\ 0 & d & f & K_{T1} & J_G \\ c & e & 0 & g & K_{T2} \end{bmatrix}$$

Besides, whatever inferior planets, the topological structures of the reconstruction designs by Wright are the same. The differences are the teeth number of the gear trains.

Regarding the structure analysis of mechanisms to generate inferior planetary motions, the mechanism for Venus' motion is presented as an example in what follows, Fig. 4.5. Based on the existing design, the characteristics of the topological structures for the interior planetary motions are concluded as follows:

1. It is a planar 6-bar mechanism with 9 joints and 1 degree of freedom.
2. It has a ground link (member 1, K_F), an input link (member 2, K_I), an output link (member 3, K_O), and 3 transmission links (members 4, 5, and 6; K_{T1}, K_{T2}, and K_{T3}).
3. It has 5 revolute joints (joints *a*, *b*, *d*, *e*, and *f*; J_R), 3 gear joints (joints *c*, *h*, and *i*; J_G), and 1 pin-in-slot joint (joint *g*; J_A).
4. It is an epicyclic gear train and formed exclusively by external gears.
5. The input link, as the carrier of the epicyclic gear train, rotates once a year.

The topological matrix of the inferior planetary mechanisms (M_T) is as follows:

$$M_T = \begin{bmatrix} K_F & J_R & J_R & J_G & 0 & 0 \\ a & K_I & 0 & J_R & J_R & J_R \\ b & 0 & K_O & 0 & 0 & J_A \\ c & d & 0 & K_{T1} & J_G & 0 \\ 0 & e & 0 & h & K_{T2} & J_G \\ 0 & f & g & 0 & i & K_{T3} \end{bmatrix}$$

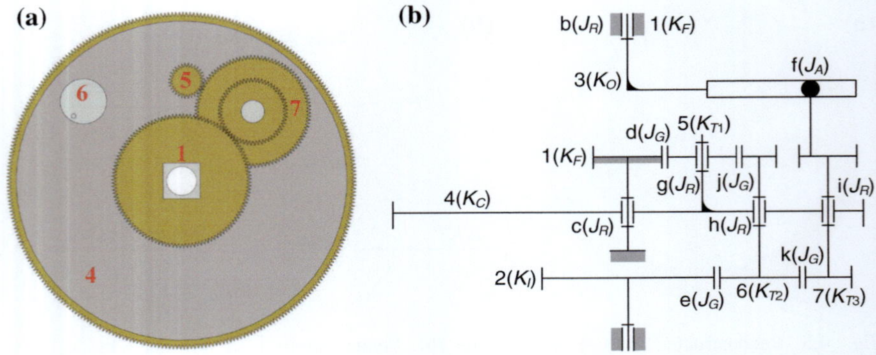

Fig. 4.6 Reconstructed partial mechanism for Jupiter's motion. **a** Physical model [13]. **b** Schematic drawing

In the same way, only one of the superior planets is presented for the structure analysis. Here, the mechanism to generate Jupiter's motion is illustrated, as shown in Fig. 4.6. The characteristics of the topological structures are concluded as follows:

1. It is a planar 7-bar mechanism with 10 joints and 1 degree of freedom.
2. It has a ground link (member 1, K_F), an input link (member 2, K_I), an output link (member 3, K_O), a carrier (member 4, K_C), and 3 transmission links (members 5, 6, and 7; K_{T1}, K_{T2}, and K_{T3}).
3. It has 6 revolute joints (joints *a*, *b*, *c*, *d*, *e*, and *f*; J_R), 2 gear joints (joints *g*, *h*, and *i*; J_G), and 1 pin-in-slot joint (joint *j*; J_A).
4. It is an epicyclic gear train and formed exclusively by external gears.
5. The transmission link, which is incident to the pin-in-slot joint, is referred as the epicycle of the epicyclic system and must rotate once a year.

The topological matrix of the superior planetary mechanisms (M_T) is as follows:

$$M_T = \begin{bmatrix} K_F & J_R & J_R & J_R & J_G & 0 & 0 \\ a & K_I & 0 & 0 & 0 & J_G & 0 \\ b & 0 & K_O & 0 & 0 & 0 & J_A \\ c & 0 & 0 & K_C & J_R & J_R & J_R \\ d & 0 & 0 & g & K_{T1} & J_G & 0 \\ 0 & e & 0 & h & j & K_{T2} & J_G \\ 0 & 0 & f & i & 0 & k & K_{T3} \end{bmatrix}$$

Wright also built the above reconstruction model containing mechanisms for the front dial. He collaborated with AG. Bromley to undertake the decoding research of the original fragments. They confirmed that some parts of the Price's design were mistaken, including the explanations of the upper back dial and the identification of the gears to transmit the power from the drive wheel to the differential gear train. Finally, based on the detection records from the cooperation with

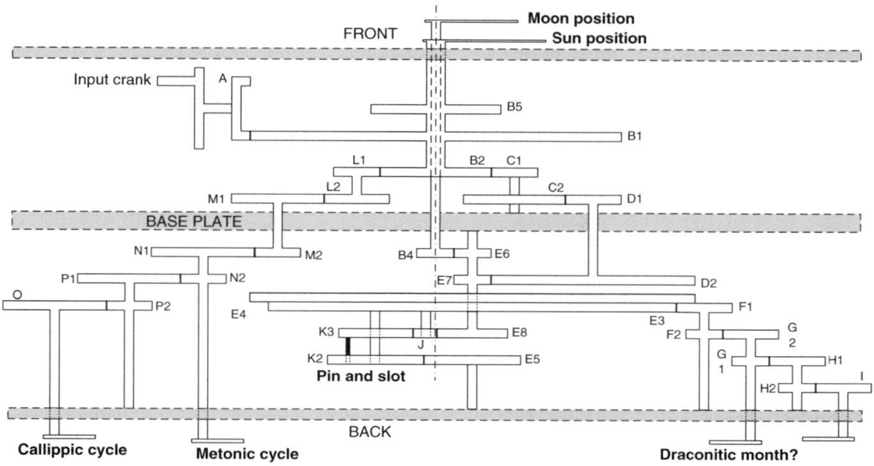

Fig. 4.7 Interior reconstruction design by Wright [15, 16]

Bromley, Wright presented the second reconstruction design with the systematic exterior appearance and interior mechanism by a series of studies, as shown in Fig. 4.7 [9, 14–18]. This reconstruction design shows new interpretations different from Price's design.

In Wright's reconstruction design, some parts of the interior structure and the types of the back dials were modified. The spiral back dials replaced the original concentric dials. The upper back dial and the lower back dial had 5-turn spiral and 4-turn spiral, respectively. Furthermore, the upper back dial and its subsidiary dial were identified for the Metonic cycle and the Callippic cycle. This viewpoint of reconstruction design is the same as one of Price's suggestions for the reconstruction design of the upper back dials. Although Price rejected such a reconstruction design for the upper back dial, Wright proved that the design concept is correct later by the measurement of advanced image-processing technology. Since the application of the lower back dials was still uncertain, Wright suggested that the lower dials are used to indicate some information related to draconitic month.

To compare with the interior structure, there are several modifications between Price's design shown in Fig. 4.1b and Wright's design shown in Fig. 4.7. The gear mechanism corresponding to the known upper back dials was reconstructed in Wright's reconstruction design. It is an ordinary gear train with 4 links, 3 revolute joints, and 2 gear joints. Besides, the gear train with 2 degrees of freedom by Price was modified as an epicyclic gear train with 6 links, 5 revolute joints, 3 gear joints, 1 pin-in-slot joint, and 1 degree of freedom, as shown in Fig. 4.7. The discovery of the pin-in-slot joint was critically important for the operation of the Wright's reconstruction structure. The two input powers to this partial mechanism were never necessary in this epicyclic gear train. An idle gear was applied to adjust the rotational direction of the output of the gear train. Thus, the purpose of the partial mechanism was changed. The lunar motion was not generated by the partial mechanism. In addition, the path of power transmission from the input to the output to generate the lunar motion is *A*–*B*1–*B*2–*C*1–*C*2–*D*1–*D*2–*E*7–*E*6–*B*4. The resulting lunar motion was demonstrated with a mean rate. The epicyclic gear train was used to generate an appropriate rate to operate the gear mechanism for the lower back dials. The topological structure of the gear mechanism for the lower back dials was not modified. But the teeth were recounted to satisfy his reconstructed lower back dials for some display functions about draconitic month.

According to Price's design shown in Fig. 4.1b, the drive wheel must turn counterclockwise so as to generate the clockwise display of the lunar motion over the front dial. And, a gear opposite to the drive wheel for the solar motion was hypothetical. Such an arrangement was overturned in the Wright's design, in which gear *B*1, i.e., the drive wheel of the Price's design, turned clockwise, not counterclockwise. The hypothetical gear opposite to gear B1 for the solar motion was not necessary. As mentioned above, the mechanism for the solar motion was generated by an epicyclic gear train existed in his reconstruction model of the front dial.

Furthermore, Wright added a new feature to the front dial of his reconstruction design in 2005 [12, 19]. This feature appeared at the top of the Antikythera device. It could demonstrate the moon phase by turning a black-and-white ball in a hole. The reconstruction work about the feature was based on the evidence from the original fragment *C*. Figure 4.8 shows the relevant excavation and its reconstruction model about the feature of the moon phase display. Such a design was nonexistent in Price's design.

Wright's research indicated that a pin-in-slot joint and a stepped stud were identified definitely from the original fragments [16]. They existed in the epicyclic gear train of the back structure. Two mobiles turned about separate centers on the stepped stud results in the eccentric distance. In addition, the pin-in-slot joint makes these two mobiles turn together. Such a design is also found in the dialwork of de' Dondi's astronomical clock modeling the Ptolemy's planetary theory. Wright understood that the pin-in-slot device can generate a variable velocity and was intended to model the anomaly motions of astronomy. It seemed to be the demonstration of Hipparchus' lunar theory. Furthermore, this discovery resulted in the rotational motion with the periodically variable speed which led to the concept of solid reconstruction design. The resulting motions satisfied the development of the astronomy at the time of Antikythera device. As a matter of fact, before the proposition of the pin-in-slot joint, Wright had applied such a joint to build a physical model of Antikythera device that displayed the anomaly motions of the Sun, the Moon, and the five planets. However, the application of the pin-in-slot joint in the epicyclic gear train was an unresolved problem in his reconstruction design.

While the pin-in-slot device introduced a roughly sinusoidal wave into the velocity ratio of the carrier of the epicyclic gear train, i.e., gears *E*3 and *E*4 in Fig. 4.7, it introduced anomaly rates to the ordinary gear train for the lower back dials. And, it seemed to be more difficult to explain the purpose of the lower back dials.

4.5 Reconstruction Work by Freeth et al.

For the Antikythera device, its third reconstruction design with the corresponding interior mechanism, as shown in Fig. 4.9, was delivered by Freeth et al. in the Antikythera Mechanism Research Project, in short AMRP. This project was an international collaboration of academia and supported simultaneously by some high image-processing technology companies. The researchers came from the academia teams, the team of the National Archaeological Museum of Athens, and the technical collaborators. In addition, T. Freeth, M. Edmunds, J. Seiradakis, Y. Bitsakis, and A.Tselikas were the core of the academic teams [2, 22–25].

Before forming the AMRP, Freeth had studied the mechanism [26–28]. In 2005, Freeth et al. firstly took effort to detect the unknown elements of excavation by the innovative image technologies and computed tomography. They collected

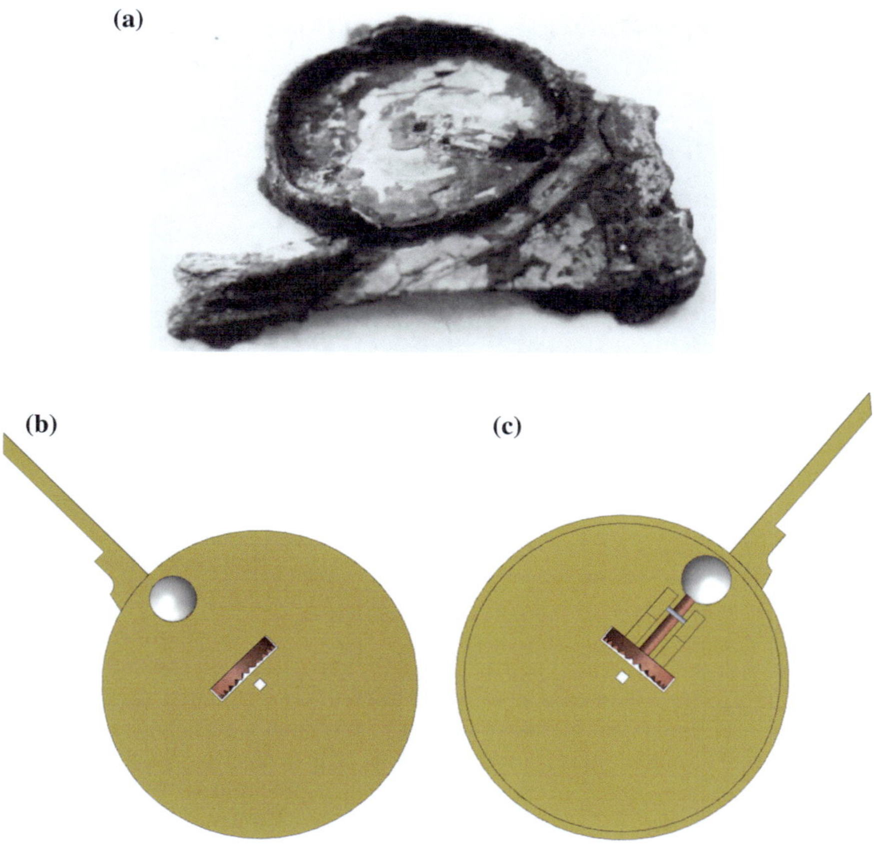

Fig. 4.8 Excavated evidence of the moon phase display device. **a** Excavation, reprinted from Ref. [9], copyright 2015, with permission from National Archaeological Museum of Aegean. **b** Top view [19]. **c** Bottom view [19]

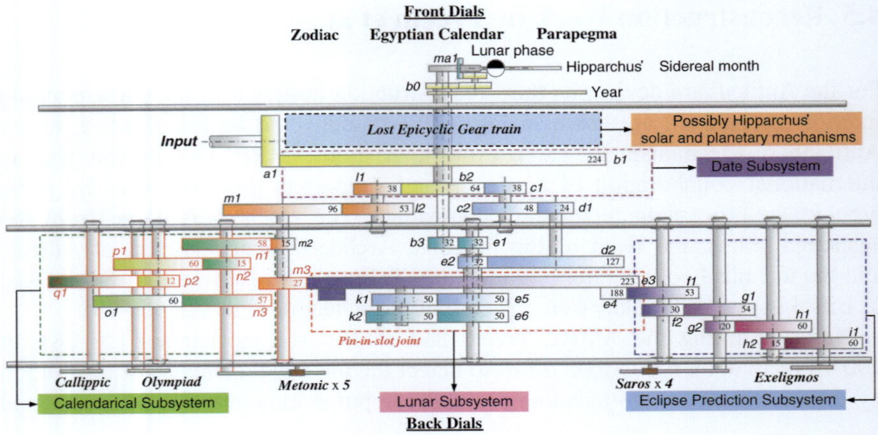

Fig. 4.9 Interior reconstruction design by Freeth et al. in 2008 [25]

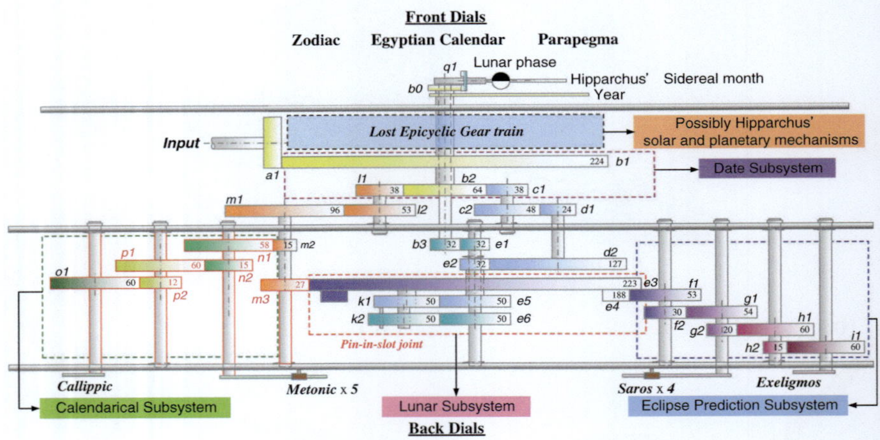

Fig. 4.10 Interior reconstruction design by Freeth et al. in 2006 [23]

the detection records to build the database of the reconstruction research. In 2006, this research team presented a new reconstruction design, based on the decoding of the inscriptions and the identification of the interior mechanism, as shown in Fig. 4.10. They modified the epicyclic gear train with a pin-in-slot joint and resolved the existing problem regarding the pin-in-slot joint in Wright's design according to their detailed inspection. The idle gear *J* shown in Wright's design, Fig. 4.7, was not existed. Gears *e*5, *e*6, *k*1, and *k*2 have similar gear sizes. Gears *e*2 and *e*5 are mounted by a hollow arbor through gear *e*3 which is a mobile on the hollow arbor. Gears *e*6 and *e*1 are mounted by a solid shaft through the hollow arbor. Gears $k1$ and $k2$ are still the mobiles on the stepped stud mounted on gear $e3$, and they turn together by a pin-in-slot joint. The modified gear mechanism should be an epicyclic gear train with 6 links, 5 revolute joints, 2 gear joints, and 1 pin-in-slot joint. This epicyclic gear train required two input controls. The transmission path of one input is $a1$–$b1$–$b2$–$c1$–$c2$–$d1$–$d2$–$e2$–$e5$, and the path of other input is $a1$–$b1$–$b2$–$l1$–$l2$–$m1$–$m2$–$m3$–$e3$. Gear $m3$ is hypothetical to generate the second input power. Such a design of the epicyclic gear train is generated based on the Hipparchus' lunar theory. Therefore, the AMRP confirmed that this epicyclic gear train must demonstrate the lunar anomaly motion.

Meanwhile, the building of the epicyclic gear train for the lunar motion affected the decoding of the lower back dials. The applications and the gear mechanisms of the lower back dials were resolved completely. The lower major dial was composed of a 4-turn spiral. The AMRP built 223 divisions in the 4-turn spiral and confirmed that the lower back dial was to show the Saros cycle. The lower subsidiary dial with 3 divisions was for the Exeligmos cycle. The corresponding gear mechanism is an ordinary gear train with 5 links, 4 revolute joints, and 3 gear joints. Gear $e4$ mounted on gear $e3$ like a ring engaged gear $f1$ to drive this gear train. The teeth were counted according to the surviving evidence to achieve these specific gear ratios for the lower back dials. Besides, the AMRP confirmed that the upper dial had 235 divisions in its 5-turn spiral to demonstrate 235-synodic months of 19-year Metonic cycle. The subsidiary dial inside the upper spiral dial with 4 divisions was for 76-year Callippic cycle. Every time the hand ran around the spiral dial 4 times, the hand of the subsidiary dial finished one time. The corresponding gear mechanism is reconstructed as an ordinary gear train with 5 links, 4 revolute joints, and 3 gear joints. Most gears of the gear mechanism were hypothetical by the AMRP in order to satisfy the Metonic cycle and the Callippic cycle. Therefore, the AMRP defined that the upper back dials were to show lunisolar calendar, and the lower back dials were used to predict the time of eclipses.

At the same time, AMRP's research showed a possible reconstruction model of the indicators for the back spiral dials, as shown in Fig. 4.11. The indicator contained a pointer and a follower with a pin. The follower could slide along the pointer, and its pin was constrained to follow the groove of the spiral dial. Thus, the follower could change its rotational radius with the rotation of the pointer driven by the output shaft of the interior gear train so as to achieve the spiral motion path.

In 2008, the AMRP presented another modified reconstruction design further, as shown in Fig. 4.9 [24, 25]. Focusing on the upper back dials, a new subsidiary dial was added and the purposes of these two upper subsidiary dials were defined again as shown in Fig. 3.6a. The right subsidiary dial, which was regarded as the Callippic cycle previously, was proved to serve as the 4-year Olympiad cycle after decoding its inscriptions. The left subsidiary dial was added and supposed to be the Callippic cycle. Most inscriptions on the spiral dial for the Metonic cycle and the subsidiary dial for the Olympiad cycle were decoded. This outstanding contribution is a powerful confirmation for the back dials. Furthermore, as the redefinition of the upper back dials, the gear mechanism related to these dials was

Fig. 4.11 A feasible pointer design of the back spiral dials [23]

modified. The ordinary gear train should have 6 links, 5 revolute joints, and 4 gear joints. Except for this modification, the remaining parts of the interior mechanism were the same as their previous reconstruction design. The latest AMRP reconstruction design shown in Fig. 4.9 is the most accepted design. Their amazing research achievements are based on the highly developed technology of computer tomography (CT). Its advanced ability of detection provides images of high resolutions of the excavated fragments to decode the above information so as to identify the mechanical elements and rebuild corresponding mechanisms. It is known that the mechanisms for the demonstration of the planetary and solar motions are not proposed, and they are regarded as the lost epicyclic gear train. Furthermore, the identified cycles are based on the calendars and astronomical cycles mentioned in Sect. 1.2. The lunar motion satisfies Hipparchus lunar theory in Sect. 1.3.2.

In 2012, a further design for the solar motion and the planetary motions by Freeth and Jones was built [29]. The relevant gear mechanisms were added in the front parts of the AMRP's reconstruction design that had been regarded as the lost epicyclic gear trains. This reconstruction research depended on his advanced identification of the inspection records of the original fragments and experience from the reconstruction of the lunar epicyclic gear train. It is hypothesized that each of epicyclic gear trains for different planets had a pin-in-slot joint. In addition, all of them were built on the surviving gear *b*1 that has 224 teeth and engages with the contrate gear connecting the input crank.

4.6 Others' Research After AD 2000

A great quantity of research work decoded successfully several mysteries of the Antikythera device and built quite a complex reconstruction model that extremely exceeded the formerly known standards of science and technology at around 150 BC. The foregoing studies build upon one another, thus providing an important basis for further research. Furthermore, more and more scholars have made their efforts to reconstruct, decode, and analyze the mechanism.

In 2008, T. Koetsier presented the historical development of reconstructing the Antikythera device containing the analysis of each previous reconstruction design and the relevant investigation of the ancient astronomy [30]. Furthermore, from the viewpoint of modern technology, C. Rossi et al. presented the study of the members of alloy of gears and the shape of teeth. According to the finding of a gear dated to the third century BC, it was confirmed that such manufacturing technique of the sophisticated gear was developed highly before the Antikythera device [31]. The excavated gear is the interesting gear of Olbia found in 2006. It is believed that the gear diameter is 43 mm and the teeth number is 55. The most important is that the teeth shape is similar to involute according to the reconstruction research. It makes the gear similar to the one in modern times. Since the excavated gear is so incomplete, the application and appearance of the original design containing this gear are unknown.

Since 2010, a fresh viewpoint regarding how the Antikythera device displays the solar and the planetary anomaly motions on the front plate was presented by Evans, Carman, and Thorndike [32–34]. Their reconstruction research revealed a special illustration of the front concentric dial to demonstrate the solar motion and follow the date of the Egyptian calendar, as shown in Fig. 4.12. The scale of the

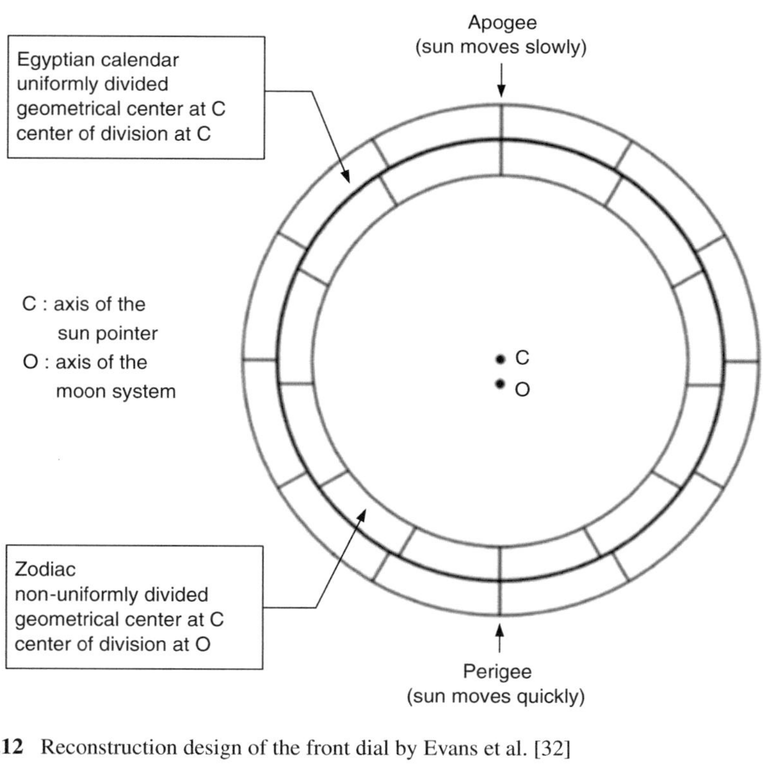

Fig. 4.12 Reconstruction design of the front dial by Evans et al. [32]

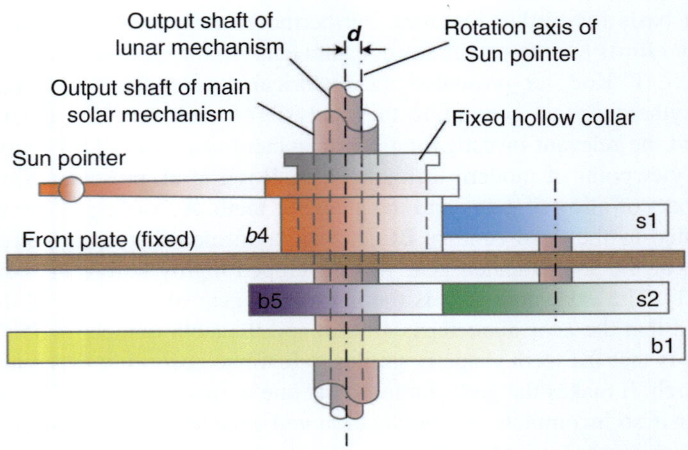

Fig. 4.13 Set of the pointers over the front dial by Evans et al. [32]

front zodiac dial (inner dial) was divided by means of a protractor placed in the off-center *C*, not in its geometric center *O*. The scale of the zodiac dial could be uniformly divided to generate the nonuniform divisions. The Egyptian calendar dial (the outer dial) was still divided in accordance with its geometric center *O*. As to the set of pointers, the Moon pointer and the date pointer were aligned with the geometric center *C* of the front concentric dial. But the Sun pointer was eccentrically set at off-center *C*, as shown in Fig. 4.13. Thus, the Sun pointer could display with a uniform rate on the zodiac dial with nonuniform scale to demonstrate the solar anomaly. As to the five planets, they were displayed, respectively, on five small dials inside the concentric dial, and their corresponding gear mechanisms were built on the main gear, as shown in Fig. 4.14. Afterward, the reconstruction design of the five planets was modified as epicyclic gear trains with a pin-in-slot joint based on the similar design concept of the lunar subsystem. These hypothetical epicyclic gear trains are established on the existing big gear *b*1 with 224 teeth. In addition, their output axles should be coaxial with the output axle of solar motion. Evans et al. presented several possible designs for a superior planet. Figure 4.15 shows the simple reconstructed gear mechanism for Mars' motion [33].

In F. Sorge's research [35], the gearing behavior of the mechanism is characterized by the knowledge of modern mechanism, and the power loss is evaluated by considering tooth collision and variable speed ratios, in order to verify the extraordinary technology. Beside, Efstathiou et al. [36] focused on perfectly reappearing the sophisticated mechanism. The geometrical dimensions of the surviving gears are investigated to accurately determine the distances between the axes of meshing gears so as to build the working model based on the existing design by Freeth et al. in 2008, as shown in Fig. 4.16. This reconstruction model is exhibited in the National Archaeological Museum of Athens. In fact, except for this model, there are three physical models of Antikythera device that were reconstructed in different times. All of them are completely different owing to the different decoding at their time periods of building.

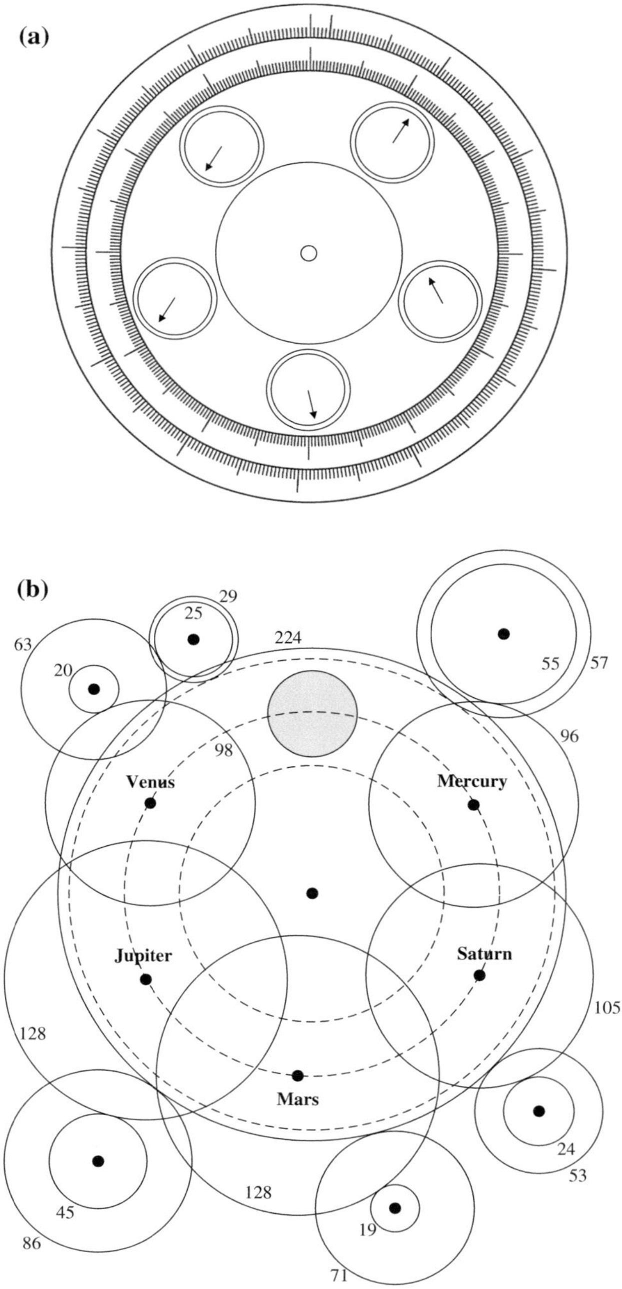

Fig. 4.14 Hypothetical reconstruction design for the motions of the five planets by Evans et al. [32]. **a** Dials of front plate. **b** Corresponding gear mechanism

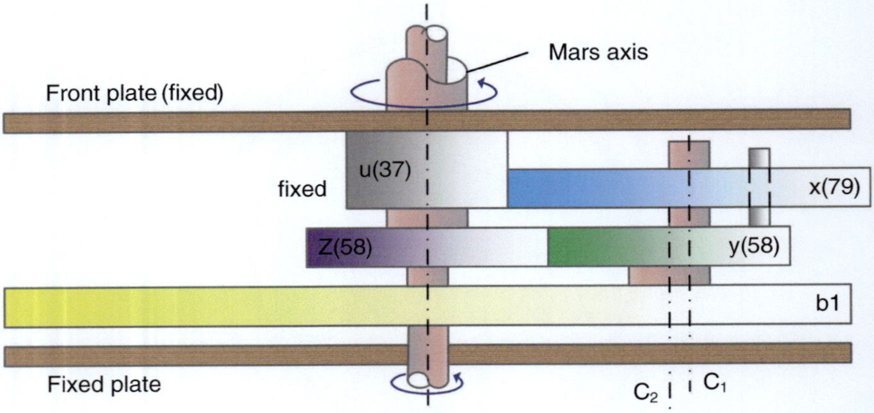

Fig. 4.15 Reconstructed gear mechanism for Mars' motion by Evans et al. [33]

Fig. 4.16 Reconstructed physical model by Efstathiou et al. reprinted from Ref. [36], copyright 2015, with permission from Elsevier

Except for academic tread in archeology and history, the surprising research results attract more attentions on the reconstruction research of the Antikythera device. In 2008, J. Marchant, a freelance journalist, introduced a full story questing for the truth of the Antikythera device through the interview with those who studied or investigated this device before [1].

In short, all of the foregoing studies show several continuous and possible revolutions for restoring the Antikythera astronomical device and establish an important basis for further investigation.

References

1. Marchant J (2008) Decoding the heavens—solving the mystery of the world's first computer. William Heinemann, Great Britain
2. Databased of the Antikythera Mechanism (2014) The Antikythera Mechanism Research Project. Available in the website. http://www.antikythera-mechanism.gr. Accessed July 2014
3. Kaltsas N, Vlachogianni E, Bouyia P (2012) The Antikythera shipwreck: the ship, the treasures, the mechanism. Hellenic Ministry of Culture and Tourism, National Archaeological Museum, Athens
4. de Price DS (1974) Gears from the Greeks: The Antikythera mechanism—a calendar computer from ca. 80 BC. Science History Publications, New York
5. de Price DS (1959) An ancient Greek computer. Sci Am 60–67
6. M. Edmunds, P. Morgan (2000) The Antikythera mechanism: still a Mystery of Greek astronomy. Astron Geophys 41:6.10–6.17
7. Field JV, Wright MT (1985) Gears from the Byzantines: a portable sundial with calendrical gearing. Ann Sci 42:87–138
8. Wright MT (2007) The Antikythera mechanism reconsidered. Interdisciplinary Sci Rev 32(1):27–43
9. Wright MT (2004) The scholar, the mechanic and the Antikythera Mechanism. Bull Sci Instrum Soc 80:4–11
10. Wright MT (2002) In the steps of the master mechanic. In: Ancient Greece and the modern world. Proceedings of a conference in ancient Olympia, 2002, University of Patras, Patras, Greece, pp 86–97
11. Wright MT (2012) The front dial of the Antikythera mechanism. In: Proceedings of the HMM2012. Springer, Netherlands, No. 15, 279–292. HMM 2012: International Symposium on History of Machines and Mechanisms, VU University Amsterdam, The Netherlands
12. Wright MT, Bromley AG (2003) Towards a new reconstruction of the Antikythera mechanism. In: Paipetis SA (eds) Proceedings of conference extraordinary machines and structures in antiquity, Peri Technon, Patras, pp 81–94. Conf. Extraordinary Machines and Structures in Antiquity, Ancient Olympia, August 2001
13. Wright MT (2002) A planetarium display for the Antikythera mechanism. Horological J 144(5):169–173
14. Wright MT (2003) Epicyclic gearing and the Antikythera mechanism: part 1. Antiquarian Horology 27:270–279
15. Wright MT (2005) The Antikythera mechanism: a new gearing scheme. Bull Sci Instrument Soc 85:2–7
16. Wright MT (2005) Epicyclic gearing and the Antikythera mechanism: part 2. Antiquarian Horology 29:51–63
17. Wright MT (2005) Counting months and years: the upper back dial of the Antikythera mechanism. Bull Sci Instrument Soc 87:8–13
18. Databasw of Wright's Model (2014) Mr. Michael Wright M.A., M.Sc., Cert. Ed., D.I.C., F.S.A. In: Centre for the History of Science, Technology and Medicine. Imperial College London. Available in the website http://www3.imperial.ac.uk/historyofscience/chostmpeople/wright. Accessed July 2014
19. Wright MT (2006) The Antikythera mechanism and the early history of the moon phase display. Antiquarian Horology 29(3):319–329
20. Wright MT (2006) Understanding the Antikythera mechanism. In: Proceedings of the 2nd international conference Ancient Greek technology, Athens, Oct 2005, pp 49–60
21. Wright MT (2013) The Antikythera mechanism: compound gear-trains for planetary indications. J Almagest 4(2)
22. Freeth T et al (2006) Decoding the Ancient Greek astronomical calculator known as the Antikythera mechanism. Nature 444:587–891 (Supplementary notes)

23. Freeth T et al (2006) Decoding the Ancient Greek astronomical calculator known as the Antikythera mechanism. Nature 444:587–891
24. Freeth T et al (2008) Calendars with olympiad display and eclipse prediction on the Antikythera mechanism. Nature 454:614–617
25. Freeth T et al (2008) Calendars with olympiad display and eclipse prediction on the Antikythera mechanism. Nature 454:614–617 (Supplementary notes)
26. Freeth T (2009) Decoding an ancient computer. Sci Am 301(6):79–83
27. Freeth T (2002) The Antikythera Mechanism: 1. Challenging the Classic Research. Mediterr Archaeol Archaeometry 2:21–35
28. Freeth T (2002) The Antikythera mechanism: 2. Is it Posidonius Orrery. Mediterr Archaeol Archaeometry 2:45–58
29. Freeth T, Jones A (2012) The Cosmos in the Antikythera mechanism. In: Institute for the study of the ancient world, New York University. Available in the website: Institute for the Study of the Ancient World (ISAW). http://dlib.nyu.edu/awdl/isaw/isaw-papers/4/. Accessed July 2014
30. Kotsier T (2008) Phase in the unraveling of the secrets of the gear system of the Antikythera Mechanism. In: Yan HS, Ceccarelli M (eds) Proceedings of HMM 2008, International Symposium on History of Machines and Mechanisms, Tainan, 2008, Springer, Netherlands, pp 269–294
31. Rossi C et al (2009) Chapter 4: ancient computation devices. In: Ceccarelli M (eds) Ancient engineers' inventions: precursors of the present. History of Mechanism and Machine Science, vol 8. Springer, Netherlands, pp 41–59
32. Evans J, Carman CC, Thorndike AS (2010) Solar Anomaly and planetary displays in the Antikythera Mechanism. J Hist Astron 42:1–39
33. Carman CC, Thorndike AS, Evans J (2012) On the pin-and-slot device of the Antikythera mechanism, with a new application to the superior planets. J Hist Astron 43(150):93–116
34. Evans J, Carman CC (2014) Mechanical astronomy: a route to the ancient discovery of epicycles and eccentrics. In: Sidoli N, Van Brummelen G (eds) Alexandria, through Baghdad: surveys and studies in the Ancient Greek and medieval islamic mathematical sciences in honor of J.L. Berggren. Springer, Berlin, pp 145–171
35. Sorge F (2012) Coupling mechanics of Antikythera gearwheels. ASME transactions. J Mech Des 134(7):061007
36. Efstathiou K, Basiakoulis A, Efstathiou M, Anastatsiou M, Seiradakis JH (2012) Determination of the gears geometrical parameters necessary for the construction of an operational model of the Antikythera mechanism. Mech Mach Theory 52:219–231
37. Clemens M (2013) The Antikythera Mechanism. In: Ancient Aliens. Available in the website. http://www.paranormalpeopleonline.com/theantikythera-mechanism/. Accessed July 2014

Chapter 5
Reconstruction Design Methodology

Reconstruction of a lost ancient mechanism is a difficult task owing to its insufficient historical records (words or pictures) or excavations that were lost, incomplete, unclear, or damaged. Such a work is usually an interdisciplinary investigation and has several possible resolutions. Decoding the mechanisms inside the Antikythera astronomical device is a typical example.

The purpose of this chapter is to introduce a design methodology for the systematic reconstruction of lost ancient mechanisms along with the resulting research achievements for reconstructing the Antikythera device by Yan and Lin. At first, the reconstruction research of lost ancient mechanisms is defined. Then, the design methodology with four steps is presented. Focusing on the research of the Antikythera device, the scopes of the historical archives are concluded. Next, the mechanism analysis of the reconstruction design by Yan and Lin is introduced. Then, the basic mechanism knowledge and the overall concepts for six interior subsystems are provided. And, the existing problems of the surviving fragments for each subsystem are described so as to define the relevant research topics. Finally, the comparisons among Yan and Lin's design and other existing designs are presented.

5.1 Reconstruction Research

The topic of reconstruction designs of lost ancient mechanisms is to rebuild the original or all feasible designs of lost mechanisms. Regarding surviving ancient mechanisms, the original designs are available in various studies. For lost ancient mechanisms, the reconstructed models are often ambiguous, and their correctness is difficult to prove. Moreover, lost ancient mechanisms have several unproven facts that should be treated as variables in reconstruction design research. As a

J.-L. Lin and H.-S. Yan, *Decoding the Mechanisms of Antikythera Astronomical Device*, DOI 10.1007/978-3-662-48447-0_5

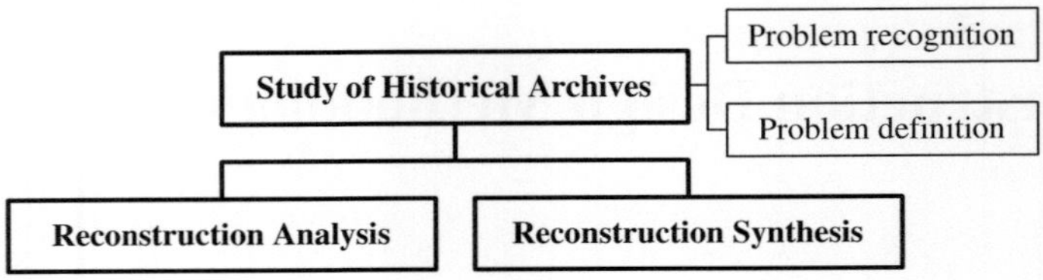

Fig. 5.1 Process of reconstruction research [1]

result, the reconstructed designs of lost ancient mechanisms may not be singular, and diverse designs are unavoidable for further proving as the original one.

The process for the reconstruction research of ancient mechanisms and machines includes the study of historical archives, reconstruction analysis, and reconstruction synthesis, as shown in Fig. 5.1 [1].

The study of historical archives is to recognize and define the problems. *Historical archives* refer to ancient manuscripts, historical artifacts, archeological data, and existing physical evidence. Ancient manuscripts are words and images found in official or unofficial historical records. Historical artifacts are buildings, implements, and paintings. Archeological data include images, language characters, symbols, and inscriptions on the archaeological discovery. In addition, the existing physical evidence contains excavated ancient devices and original historical materials. *Problem recognition* is to verify and evaluate historical archives. Although the text of ancient literature is generally simplified and may be inaccurate or exaggerated, no past technological experiences and records should be ignored. Moreover, attentions should be given to the names and terminologies of ancient mechanisms. They may have different names and definitions in different areas and eras. For example, planetarium, astrarium, and astronomical clock refer to the same mechanical device. Therefore, according to extensive collection and detail identification, useful information relevant to ancient mechanisms can be extracted. Furthermore, mechanisms in different eras were affected by their corresponding manufacturing technology and craftsmanship. Accordingly, for some mechanisms that existed in different eras or still exist today, attention should be given to the differences in design and craftsmanship of subjects' time periods. *Problem definition* is to precisely define the problem based on modern science and technology so as to provide a fresh and clear base for reconstruction research. Problem definition includes the development of design specifications and the conclusion of design constraints. In general, each era had its own terminologies or jargons regarding engineering technology and craftsmanship and may not be comprehended directly and correctly by people today. As a result, it is necessary to study the evolution of terminologies or jargons so as to understand their meanings and representative designs. Meanwhile, the design, construction, technology, and craftsmanship of ancient mechanisms should be investigated. Thus, the historical archives of ancient mechanisms could be decoded and transformed into modern knowledge and information.

Reconstruction analysis and *reconstruction synthesis* are regarded to as two parts of the reconstruction design. For surviving ancient mechanisms, their task of reconstruction design is mainly on reconstruction analysis. Its purpose is to study, test, and verify the proven subjects by modern engineering technology and methods. Surviving ancient mechanisms can be seen in substantial objects. Their excavations provide source materials for various modern analytical methods, such as technical testing, kinematic analysis, force analysis, engineering drafting, statistical analysis, simulation variation, and scientific deductions. Modern equipments are used to carry out accurate measurement and testing regarding the geometric parameters of all members of excavations. Building simulation models according to the measured data could reappear the ancient mechanisms so as to experiment on their ability of operation. It is also helpful for detecting the metallographic structures to determine the composition of alloys. In fact, these analytical results of surviving ancient mechanisms contributed a lot to the reconstruction research of other ancient mechanisms in the same era, yet without surviving evidence.

For lost ancient mechanisms, their main task of reconstruction research is on reconstruction synthesis. Its purpose is to regenerate ancient mechanisms that are consistent with historical records, excavated evidence, decoded information, and the levels of ancient technology and craftsmanship subject to the developed design specifications. Reconstruction synthesis is a research that has been practiced in the long-past eras. Most efforts of reconstruction synthesis were based on verification of historical records and available artifacts, and knowledge, experiences, and judgments of scholars. However, very few scholars focused on studying lost ancient mechanisms based on a systematic academic approach.

5.2 Reconstruction Design Methodology

A methodology, which is a set of procedures or design steps, is able to systematically solve a defined problem. It includes a body of methods, rules, and postulates. In the past several decades, available methodologies for the structural synthesis of mechanical devices can be divided into four types. The first one is the design method based on the structure of mechanisms. The second is the design method based on the concept of modularization. The third is based on evolutionary perspective. The fourth is the design method based on available database and experience. These methods have their own advantages and disadvantages and can be used interchangeably without conflicts.

The reconstruction designs of lost ancient mechanical devices focused on the structural synthesis for rebuilding the original or the most possible designs. Feasible results satisfying the historical background and the standard of science and technology at the subject's time period might be numerous owing to the incomplete and lacking of historical information. Here, a methodology to define relevant problems, identify appropriate structural characteristics, and reconstruct all feasible mechanisms for the reconstruction designs of lost ancient mechanisms is introduced.

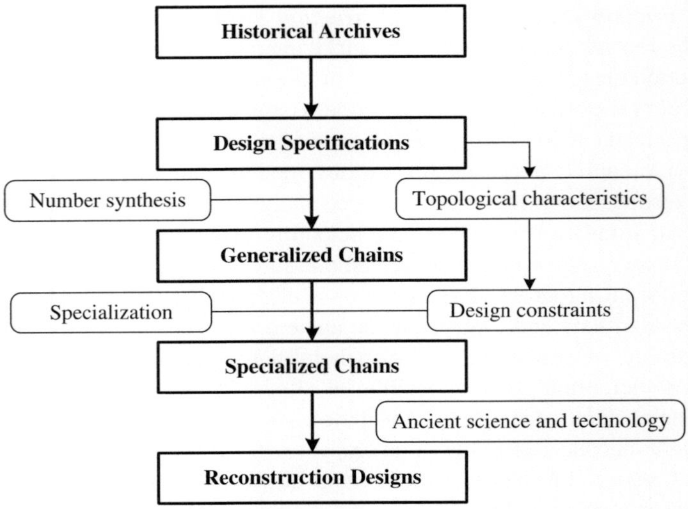

Fig. 5.2 Reconstruction design methodology of lost ancient mechanisms [1–3]

The methodology for the reconstruction designs of lost ancient mechanisms, shown in Fig. 5.2, is developed based on the methodology for the creative designs of modern mechanisms [1–3]. It includes four steps: design specifications, generalized chains, specialized chains, and reconstruction designs. Each step is, respectively, introduced in Sects. 5.2.1–5.2.4. The initial step of the methodology is to establish the historical archives based on the investigation of the relevant historical records and background. Since the research subjects contained in the step of historical archives are varied with different research topics, the applications of the design procedure are flexible, i.e., each research topic must be related to its individual research background.

The investigation of historical archives contributes to the follow-up steps of the design procedure. In this step, ancient mechanisms are redefined, and the corresponding design specifications and constraints are concluded based on modern engineering science and technology. Its purpose is to transform the problems of reconstruction research of ancient devices into modern mechanical designs based on available engineering knowledge and to resolve these problems by modern engineering design approaches. For example, a simple and humble connection method of two elements in an ancient mechanism, such as wrapping by a rope or combining by a wooden pin, can be represented with an equivalent joint defined by modern mechanisms to clearly describe its possible motions. Thus, the lost, incomplete, or unclear ancient mechanisms can be regarded to as equivalent modern mechanical designs. Moreover, the investigation works, respectively, provide important historical information for further research regarding the proven subjects and possible configuration designs of lost mechanisms.

5.2.1 Design Specifications

In the first step of the reconstruction design methodology, the "Design specifications" is to develop design specifications of the lost mechanism based on the study of available historical archives and conclude topological structure of mechanisms of available reconstruction designs or other surviving ancient mechanisms with the same functions in the subject's time period.

The design specifications of mechanisms are statements describing the numbers and the types of members and joints and the type of input of a mechanism. They should be defined at the beginning of the work and observed throughout the process of the reconstruction design. In order to recognize the design specifications, the study of available historical literatures, records, and ancient astronomical theories in advance is indispensable. Once the characteristics of topological structure of the target mechanism are identified, the specifications can be clearly defined.

5.2.2 Generalized Chains

The second step of the reconstruction design methodology "generalized chains" is to obtain the atlas of generalized chains with the required numbers of members and joints as specified in the concluded topological characteristics of the design in Sect. 5.2.1 based on the concept of number synthesis of mechanisms [2, 3].

A generalized joint is a joint in general. For the design of Antikythera device, it can be a revolute joint, a gear joint, or a pin-in-slot joint. Figure 5.3a shows a simple generalized joint. A generalized link is a link with generalized joints. For the design of Antikythera device, it can be links and gears, and it can be a binary link (a link with 2 incident joints), a ternary link (a link with 3 incident joints), a quaternary link (a link with 4 incident joints), etc. Graphically, a generalized link with N_J incident joints is symbolized by a shaded N_J-sided polygon with small circle as vertices. Figure 5.3b–d shows a binary, ternary, and quaternary generalized link, respectively.

A generalized chain consists of generalized links connected by generalized joints. It is connected, closed, without any bridge link, and with simple joints only [3]. A (N_L, N_J) generalized chain refers to a generalized chain with N_L-generalized

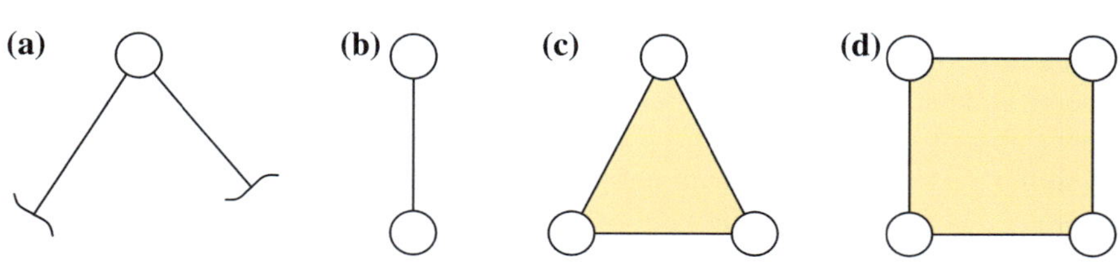

Fig. 5.3 Graphical representations of generalized joint and links

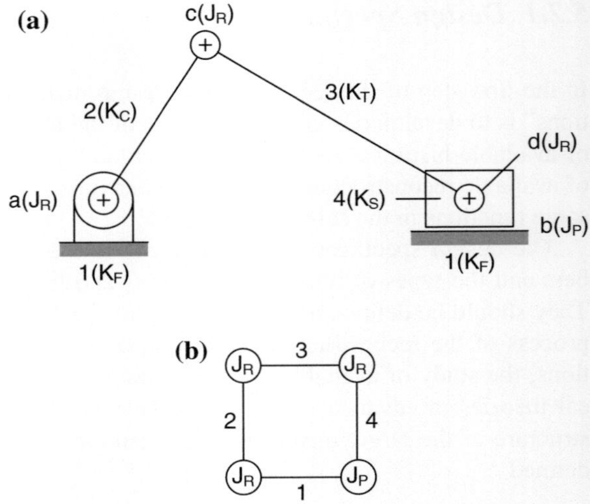

Fig. 5.4 Generalization of a crank–slider mechanism, **a** 4-bar slider–crank mechanism, **b** corresponding generalized chain

links and N_J generalized joints. For instance, Fig. 5.4a is a slider–crank mechanism. It is a 4-bar mechanism with 3 revolute joints and 1 prismatic joint. The frame (K_F) is generalized into binary link 1, the crank (K_C) is generalized into binary link 2, the connecting link (K_T) is generalized into binary link 3, and the slider (K_S) is generalized into binary link 4. The revolute joints (joints *a*, *c,* and *d*) are replaced by generalized joint with letter J_R indicating the revolute joint. And the prismatic joint (joint *b*) is replaced by a generalized joint with letter J_P indicating the prismatic joint. Therefore, the 4-bar slider–crank mechanism is generalized into its corresponding generalized chain, and it is a (4, 4) generalized chain, as shown in Fig. 5.4b. Another example is the gear mechanism of Persian astrolabe with calendrical gearing shown in Fig. 2.12. Each arbor (axis *A*–*E*) and its own gears are regarded as a link, and the frame is also a link. Therefore, the gear mechanism contains 6 links and 9 joints including 4 gear joints and 5 revolute joints. The frame is generalized into quineary link 1; arbor *A* with a gear is generalized into binary link 2; arbor *B* with two gears is generalized into ternary link 3; arbor *C* with one gear is generalized into ternary link 4; arbor *D* with two gears is generalized into ternary link 5; and arbor *E* with one gear is generalized into binary link 6. After such a process of generalization, its corresponding (6, 9) generalized chain is shown in Fig. 5.5. Therefore, a generalized chain can definitely indicate its topology characteristics, including the number and the type of links, the number of joints, and the incidence among links and joints with an abstract scheme simpler than mechanical and schematic drawings. It is a succinct graphical representation for a mechanism in mechanical engineering. These characteristics regarding the topological structure of generalized chains can be represented by the topology matrix, M_T, that is introduced in Sect. 5.4.1.

The process to transform a mechanism involving various types of members and joints into a generalized chain with only generalized links and generalized joints is

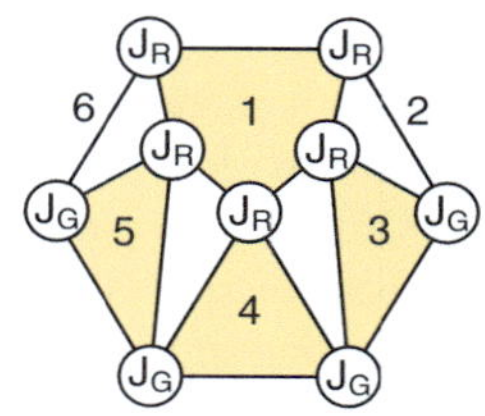

Fig. 5.5 Generalized chain corresponding to the Persian astrolabe with calendrical gearing shown in Fig. 2.12

generalization, i.e., transforming mechanical, schematic, or 3D drawings into simple graphs. This process of generalization is based on a set of generalizing rules that are derived according to defined generalizing principles [1, 3]. Furthermore, the characteristics of topological structures of a mechanism and its corresponding generalized chain are consistent.

The generalization of available designs of mechanical devices is helpful to mechanism analysis and synthesis. All possible generalized chains that have the same numbers of links and joints as the original generalized chain or the requirements of design specifications are synthesized through the concept of number synthesis which has been a subject of numerous studies in the past years. The atlases of generalized chains for the reconstruction designs of the Antikythera device are available in reference [3].

5.2.3 Specialized Chains

The third step of the reconstruction design methodology "specialized chains" is to assign the required types of members and joints in each available generalized chain, subject to the design constraints concluded from the topological characteristics of available designs and the historical archives. This process is called specialization. A generalized chain after specialization is called a specialized chain. And, a specialized chain subject to design constraints is called a feasible specialized chain

The topic of design constraints is one of the core concepts in this step to derive feasible designs from numerous mechanisms with different topological structures. In the process of reconstruction synthesis, design constraints for the identification of each link in the mechanism are based on the study of historical records and existing designs.

5.2.4 Reconstruction Designs

The final step of the reconstruction design methodology "reconstruction designs" is to particularize each specialized chain obtained in Sect. 5.2.3 into its corresponding schematic format to have the atlas of reconstruction designs that meet the science and technology standards of the subject's time period by utilizing the mechanical evolution and variation theory to perform a mechanism equivalent transformation.

Graphically, particularization is the reverse process of generalization. The atlas of feasible specialized chains could be particularized into the corresponding schematic diagrams through particularization. In order to generate all feasible reconstruction designs, the feasibility consistent with the geometry constraint of the existing evidences and the standard of the ancient technology and science of the subject's period must be considered in detail [3].

5.3 Historical Archives of Antikythera Device

The reconstruction design of Antikythera device is defined as one of the reconstruction researches of lost ancient mechanisms. The related historical archives are built through the exhaustive study of available literatures, including the decoded information, the detected evidence, the ancient astronomy, the ancient astronomical instruments, and the kinematic and mechanism analyses [4], as shown in Fig. 5.6. In what follows, each research work of historical archives is introduced.

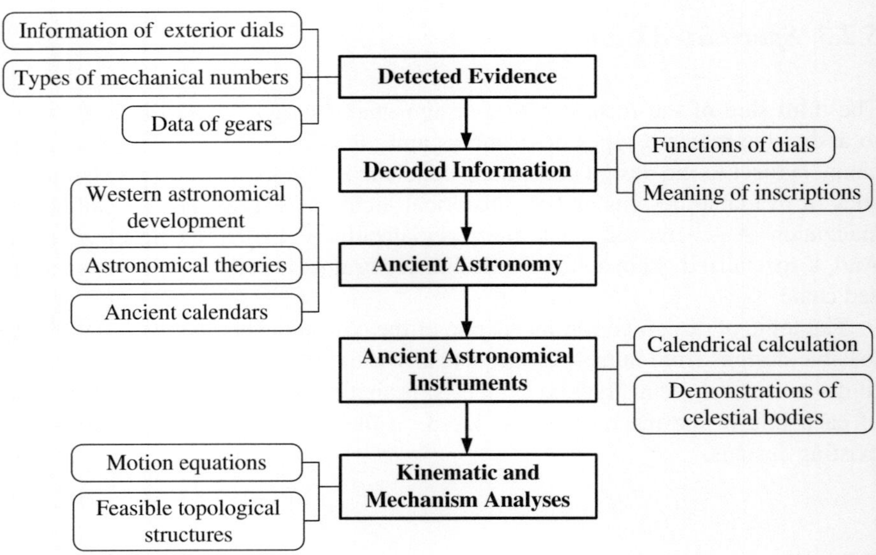

Fig. 5.6 Historical archives of the Antikythera mechanism [4]

5.3.1 Detected Evidence

The surviving wreckages of Antikythera device shown in Fig. 3.3 provide powerful knowledge in defining the fundamental design constraints and identifying the topological structure of the original mechanism, even if the excavated objects are incomplete and damaged. In previous research, a great number of inspection results are amassed with the developed image-processing technology. These results clearly indicated that Antikythera device was composed of gears and links. In addition, the specific design of pin-in-slot devices that is well known in the design and manufacturing of astronomical clocks was applied. Whether the applied mechanical members meet the science and technology subject to the given time period is undoubted. The amounts, the positions, and the styles of the exterior dials are confirmed in turn by the inspection records. Furthermore, the number of detected gears is nearly thirty to date. The corresponding teeth are also identified from the photographs of excavated fragments in related study.

5.3.2 Decoded Information

The appearance of the corroded fragments of Antikythera device was covered with copious inscriptions. Some inscriptions are observed by naked eyes, and some are obscure. For a long time, these inscriptions were detected through the technical assistance of advanced image processing and decoded gradually by scientists and archaeologists. The functions of the dials of Antikythera device are clearer and clearer with the decoding of the partial inscriptions indicating the graduations of dials. Scholars and experts are able to confirm what these exterior dials mean and how the device operates. As mentioned above, the front plate of the device can display the cyclic motions of the celestial bodies over the inner zodiac dial, including the Sun, the Moon, and the five planets. Also, it can indicate the date of the Egyptian calendar over the outer calendar dial. The back plate of the device can predict the appearances of solar and lunar eclipses and calculates the calendrical and the astronomical periods. All of these known functions are revealed by the decoding of the inscriptions.

In historical development of the reconstruction works of Antikythera device, the functions of dials kept on being corrected with the decoded information. Also, the corresponding mechanisms were modified. Take the reconstruction of the dials of the calendrical subsystem, for example, one of two subsidiary dials was identified firstly to show the Callippic period in 2006. However, with the further decoding of inscriptions, this subsidiary dial for the Callippic period was corrected for the Olympiad cycle, and another subsidiary dial was identified for the Callippic period in 2008. Therefore, it is believed that the decoded information is the chief resource to define clearly the output functions of the device for each subsystem.

All reconstruction works must begin from the decoding of inscriptions. Once the decoded information are updated or corrected in the future, the corresponding mechanisms can be reconstructed accordingly.

5.3.3 Ancient Astronomy

The Antikythera device is treated as an astronomical computer date back to more than 100 BC. All deciphered functions directly reflect the brilliant achievements on astronomy at that time. This device must be designated in accordance with the observable records of astronomy phenomena and the developed astronomical theories describing the motions of celestial bodies. Therefore, the study of astronomy subject to the time period is indispensable. For any lost subsystem, the generations of the designs with the specific functions are completely based on the investigations of ancient astronomical theories.

Indeed, the Western astronomy was developed highly on either the observation of phenomenon or the astronomical theory at the time of Antikythera device. The geocentric theory, in which the Earth is the center of the universe, was almost formed. The Sun, the Moon, and the five planets, including Venus, Mercury, Mars, Jupiter, and Saturn, move around the Earth. The most important of all is that the geometrical models by Apollonius and Hipparchus clearly illustrated the motions of the celestial bodies. Both of them are the important astronomers at the time of the Antikythera device, and their achievements should affect the development of this invention.

5.3.4 Ancient Astronomical Instruments

It is believed that the astronomical developments of astronomy are continuous and affected mutually by other civilizations. For the astronomical instruments demonstrating the astronomical theories, the relevant manufacture techniques are regarded to be employed by a traditional group of craftsmen. Logically, other ancient astronomical instruments, not focusing on Greek ones, may adopt the same or similar design concepts to achieve the required functions, even though none of them is the Antikythera device. For this reason, the study of the historical developments of the ancient astronomical instruments introduced in Chap. 2 is helpful to conclude the required design constraints.

5.3.5 Modern Kinematic and Mechanism Analyses

It is well understood in modern times that numerous mechanisms with different topological structures can generate the same output functions. Since mechanisms of calendrical and lost subsystems of Antikythera device are unclear or unknown, it is an obvious problem to conclude the appropriate design constraints from numerous types of mechanisms.

Modern kinematic and mechanism analyses provide a strong assistance to understanding the ancient mechanisms. No matter what subsystems of the Antikythera device have, it is an indispensable step in reconstruction design. Through the process of the kinematic analysis, the motion equations for specific celestial bodies can be derived. From the viewpoint of practical design, a mechanism with less number of members to generate the required functions is better. In accordance with such a logic reasoning, the possible topological structures of indefinite subsystems can be derived systematically through the modern mechanism analysis.

In summary, through the investigation of the initial four works (detected evidence, decoded information, ancient astronomy, and ancient astronomical instruments), it is known that the solar and the planetary motions are demonstrated on the front concentric dial. The kinematic characteristics of the solar and the planetary anomaly motions are concluded from ancient astronomical theories. The anomaly motions of celestial bodies are demonstrated completely by the gear mechanisms with a pin-in-slot joint. In addition, this mechanical design was proved in several ancient astronomical instruments. As a result, it is believed that such a design and technique could be preserved.

One important contribution of building historical archives of Antikythera device is to analyze how many links and joints in the design. This is essential for the lost and incomplete excavations. The concepts of using modern mechanism to resolve the difficulty are applied, i.e., the unknown numbers of mechanical elements (gears and frame) of the lost subsystems. The relations, regarding the numbers of links and joints in a feasible reconstruction design, are concluded by mechanism analysis of gear trains. Therefore, the number of mechanical elements in all designs of gear trains can be calculated. Furthermore, according to the kinematic analysis of gear trains with a pin-in-slot joint and the kinematic characteristics concluded from the ancient astronomical theories, it is concluded that the types of links, which are adjacent to each other with a pin in slot, are determined with the number of mechanical elements. In fact, all concluded information of historical archives, including design specifications and design constraints, is useful for each step of the design procedure.

5.4 Reconstruction Research by Yan and Lin

Yan and his research group have been focusing on the reconstruction designs of ancient mechanisms ever since 1990 [1]. Many lost mechanisms in ancient China were reconstructed, such as the wooden horse carriage, south-pointing chariots, Zhang Heng's seismoscope, and Su Song's escapement regulator. Furthermore, Yan and Lin started to study the reconstruction research of the Antikythera device in 2007. Based on the systematic reconstruction design methodology introduced in Sect. 5.2 and the historical archives introduced in Sect. 5.3, they synthesize all possible designs for the unclear, incomplete, and lost parts of the interior mechanisms.

In 2011, Yan and Lin presented 48 possible reconstruction designs of the whole interior mechanism, based on the systematic design methodology [4], as shown in Appendix A. These designs satisfy the surviving evidence known at present and the technology and science of the subject's time period. Figure 3.9 shows one of these 48 reconstruction designs. The interior design contains 6 subsystems mentioned in Sect. 3.3.4. In what follows, basic concepts of mechanism that are useful to the structure analysis are introduced. Then, a design is used to illustrate the structure analysis of the interior subsystems of the Antikythera device, to help understanding this complicated mechanical device. The available information, including the surviving evidence to support these subsystems, is described. And thus, the existing problems in each subsystem are described definitely.

It is understood that the existing reconstruction designs described in Chap. 4 are based on the surviving evidence of excavation, the literatures of ancient astronomy, and the professional knowledge of ancient astronomical instruments. The application of modern detection technology is necessary to obtain the detection statics from the excavation. The inscriptions are decoded, and the interior mechanisms are restored by these research bases. Nevertheless, Yan and Lin use a different way for decoding the Antikythera device. They exhaustively study the existing research to get the sufficient information of surviving evidence. They apply the professional knowledge of modern mechanical engineering to present a design methodology for resolving this research. This systematic design methodology introduced in Sect. 5.2 is an innovative tool for the reconstruction work of the Antikythera device.

5.4.1 Concepts of Mechanical Designs

In order to explain clearly the structure analysis of the known reconstruction designs, some basic concepts of modern mechanical designs for this topic are presented to avoid tediously describing the professional knowledge of mechanisms in the follow-up chapters. Here, mechanical members, joints, degrees of freedom, and topology structure used in the reconstruction designs of Antikythera device are introduced.

5.4.1.1 Mechanical Members

Mechanical members, or just members in short, are resistant bodies that collectively form mechanical devices. They can be rigid members (as kinematic links, sliders, rollers, gears, cams, power screws, shafts, keys, etc.), flexible members

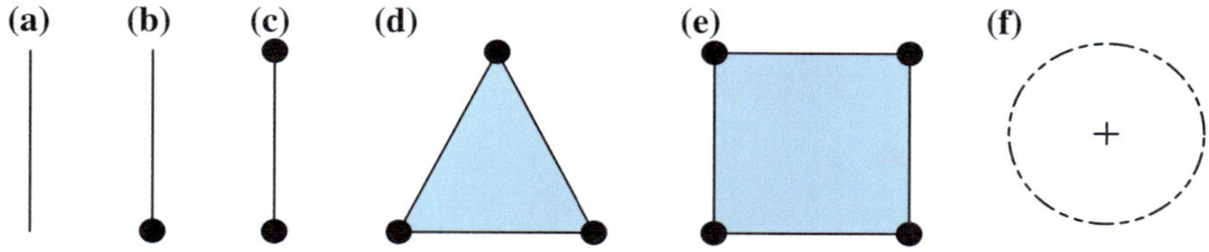

Fig. 5.7 Schemes of mechanical elements

(as springs), or compression members (as airs or fluids). The functions of those members are to provide possible relative motions with others. In what follows, kinematic links and gears are presented.

Link or Kinematic Link ($\boldsymbol{K}_{\mathrm{L}}$)

A *kinematic link* (K_{L}), or just *link* in short, is a rigid member for the use of holding its joints apart and transmitting motions and forces. Globally, any rigid mechanical member is a kinematic link. Links can be classified based on the number of incident joints. A *separated link* is one with zero incident joints. A *singular link* is one with 1 incident joint. A *binary link* is one with 2 incident joints. A *ternary link* is one with 3 incident joints. A *quaternary link* is one with 4 incident joints. And, an L_i-link is one with i incident joints. Graphically, a link with i incident joints is symbolized by a shaded, i-sided polygon with small circles on the vertices indicating incident joints. Figure 5.7a–e shows the schematic representations of a separated link, singular link, binary link, ternary link, and quaternary link, respectively [1, 3, 5].

Gear ($\boldsymbol{K}_{\mathrm{G}}$)

Gears (K_{G}) are links that are used, by means of successively engaging teeth, to provide positive motion from a rotating shaft to another that rotates, or from a rotating shaft to a member that translates. Modern gears can be classified as spur gears, bevel gears, helical gears, and worms and worm gears. Figure 5.7f shows the schematic representation of a typical gear. In the mechanisms of Antikythera device, all gears are spur gears except for two contrate (crown) gears, respectively, for the input crank and the moon phase display device.

5.4.1.2 Joints

In order to make mechanical members useful, they must be connected by certain means. That part of a member that is connected to a part of another member is called a pairing element. Two pairing elements that belong to two different

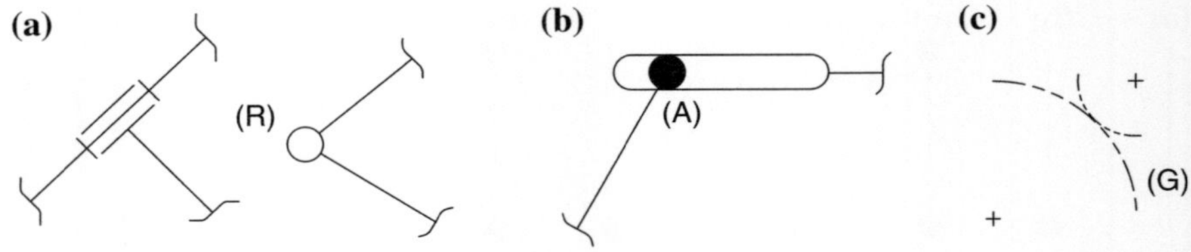

Fig. 5.8 Schemes of joints

members and are connected together form a joint or kinematic pair [1, 3, 5, 6]. In what follows, the functional descriptions of some basic joints applied in the Antikythera device are introduced and their corresponding schematic representations are shown in Fig. 5.8.

Revolute Joint ($\boldsymbol{J}_{\mathrm{R}}$)

For a *revolute joint* (J_{R}), the relative motion between two incident members is rotating about an axis. Figure 5.8a shows the schematic representation of a revolute joint.

Pin-in-Slot Joint ($\boldsymbol{J}_{\mathrm{A}}$)

For a *pin-in-slot joint* (J_{A}), the relative motion between two incident members is the combination of rolling and sliding. Figure 5.6b shows the schematic representation of a pin-in-slot joint.

Gear Joint ($\boldsymbol{J}_{\mathrm{G}}$)

For a *gear joint* (J_{G}), the relative motion between two incident members is also the combination of rolling and sliding. Figure 5.6c shows the schematic representation of a gear joint.

5.4.1.3 Degrees of Freedom

The number of degrees of freedom of a mechanism determines how many independent inputs the design must have in order to perform a useful engineering purpose. A mechanism with a positive number of degrees of freedom and with the same number of independent inputs has constrained motion [1, 3–6]. Constrained motion means that when any point on an input member of the mechanism is moved in a prescribed way, all other moving points of the mechanism have uniquely determined motions.

5.4.1.4 Topological Structure

The topological structure of a mechanism is characterized by its types and numbers of links and joints, and the incidences between them. Two mechanisms are regarded as isomorphic ones if they have the same topological structures. The isomorphism of mechanisms can be identified based on the concept of a matrix named topology matrix, which is a powerful tool for the representation of the topological structures of various kinematic chains.

The topology matrix, M_T, of an (N_L, N_J) kinematic chain or mechanism is an N_L by N_L matrix. Its diagonal element is $e_{ii} = u$ if the type of member is u. Its upper off-diagonal entry is $e_{ik} = v$ $(i < k)$ if the type of the joint incident to members i and k is v. Its lower diagonal entry is $e_{ki} = w$ if the assigned name of the joint is w. Furthermore, $e_{ik} = e_{ki} = 0$ if members i and k are not adjacent.

For instance, Fig. 5.9 is a (5, 6) epicyclic gear train in which K_F is the ground link (member 1), K_{SI} is a sun gear as the input link (member 2), K_P is a compound planet gear (member 3), K_C is a carrier (member 4), and K_{SO} is a sun gear as the output link (member 5). The joint incident to K_F and K_{SI} is a revolute joint (a, J_R), the joint incident to K_F and K_C is a revolute joint (b, J_R), the joint incident to K_F and K_{SO} is a revolute joint (c, J_R), the joint incident to K_{SI} and K_P is a gear joint (d, J_G), the joint incident to K_P and K_{SO} is a gear joint (e, J_G), and the joint incident to K_P and K_C is a revolute joint (f, J_R). Therefore, the topology matrix, M_T, of this mechanism is:

$$M_T = \begin{bmatrix} K_F & J_R & 0 & J_R & J_R \\ a & K_{SI} & J_G & 0 & 0 \\ 0 & d & K_P & J_R & J_G \\ b & 0 & e & K_C & 0 \\ c & 0 & f & 0 & K_{SO} \end{bmatrix}$$

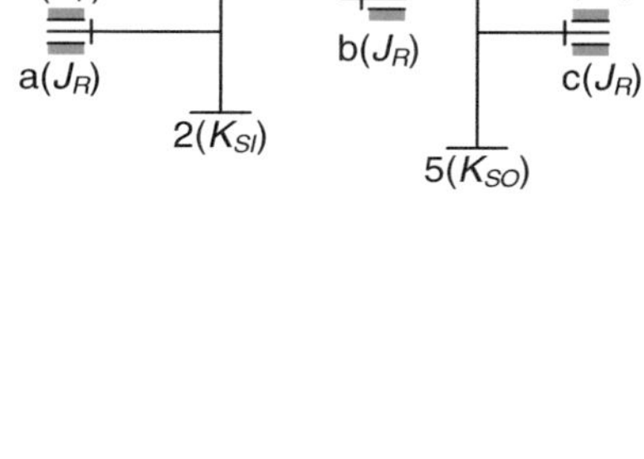

Fig. 5.9 A (5, 6) epicyclic gear train

5.4.2 Date Subsystem

The date subsystem is to operate the date pointer to generate daily motion on the front outer dial for the Egyptian calendar. Figure 5.10 shows the schematic drawing of the date subsystem. This subsystem was completely detected from the surviving fragments. The corresponding reconstruction design was presented firstly by Price [7]. Based on the surviving evidence, this subsystem contains 3 members (member 1, K_F; member 2, K_I; member 3, K_O) and 3 joints (joint a, J_R; joint b, J_R; joint c, J_G). Member 1 is the frame. Member 2 is the contrate gear connecting to the input crank and is estimated with about 48 teeth. And member 3 is a compound gear with 224 teeth and 64 teeth. The gear with 224 teeth is the biggest gear in all surviving gears.

To observe overall interior structure, the rotation of the contrate gear is also the input source of the mechanical device. All existing designs consistently indicate the characteristics of the topological structure of the date subsystem as follows:

1. The mechanism of date subsystem is a planar 3-bar mechanism with 2 revolute joints, 1 gear joint, and 1 degree of freedom.
2. The mechanism has a ground link (member 1, K_F) as frame, an input link (member 2, K_I), and an output link (member 3, K_O).
3. The mechanism has 2 revolute joints (joints a and b; J_R) and 1 gear joint (joint c; J_G).
4. The input link is the contrate gear $a1$, driven by the manual crank.
5. The output link consisting of gears $b1$ and $b2$, rotates clockwise at the rate of once in every tropical year.

The corresponding topology matrix (M_T) is:

$$M_T = \begin{bmatrix} K_F & J_R & J_R \\ a & K_I & J_G \\ b & c & K_O \end{bmatrix}$$

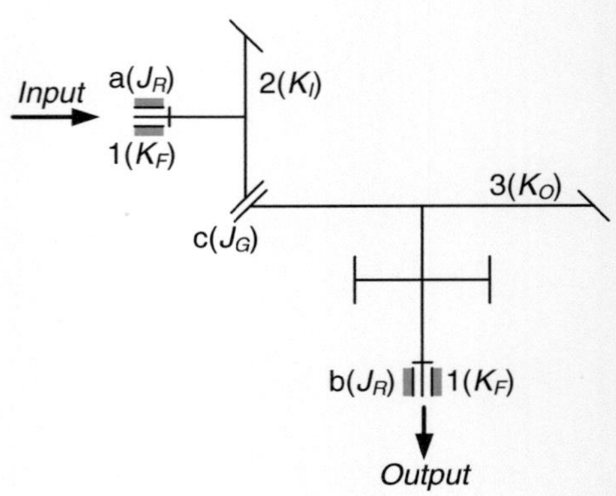

Fig. 5.10 Schematic drawing of the date subsystem

5.4.3 Eclipse Prediction Subsystem

The eclipse prediction subsystem is to predict the time of the solar and lunar eclipses. This subsystem serves for the Saros dial and the Exeligmos subsidiary dial. In addition, it is one of the subsystems which are decoded completely. All elements, including the numbers and teeth of gears, are identified. In the subsystem, there are 7 gears with 53, 30, 54, 20, 60, 15, and 60 teeth, respectively. Furthermore, these gears, respectively, position on four shafts, as shown in Fig. 3.9. Among these four existing shafts, two shafts are, respectively, with the rate of Saros cycle and the rate of the Exeligmos cycle.

Figure 5.11 shows the schematic drawing of the eclipse prediction subsystem. Its characteristics of the topological structure are concluded as follows:

1. It is a planar 5-bar mechanism with 4 revolute joints, 3 gear joints, and with 1 degree of freedom.
2. It has a ground link (member 1, K_F) as frame, an input link (member 2, K_I), a Saros cycle link (member 3, K_S), a transmission link (member 4, K_T), and an Exeligmos cycle link (member 5, K_E). The Saros cycle link and the transmission link are driven links. And the Saros cycle link shows the Saros cycle. The Exeligmos cycle link is the output link to show the Exeligmos cycle.
3. It has 4 revolute joints (joints *a*, *b*, *c*, and *d*; J_R) and 3 gear joints (joints *e*, *f*, and *g*; J_G).
4. The mechanism is an ordinary gear train, and all gears are spur gears with triangle teeth shape.
5. The input link has to engage with the existing gear (no. 49), as shown in Fig. 3.9. The Saros cycle link has to rotate clockwise at the rate of 4 turns in every 18 tropical years. The transmission link is to provide the required rate changes. And the Exeligmos cycle link must rotate clockwise at the rate of once in every 76 tropical years.

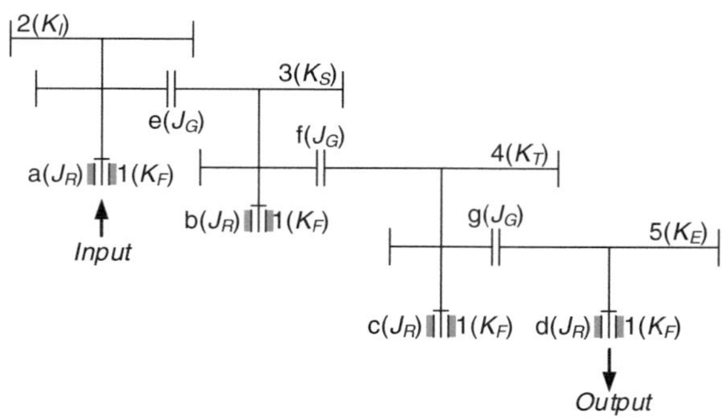

Fig. 5.11 Schematic drawing of the eclipse prediction subsystem

Its corresponding topology matrix (M_T) is:

$$M_T = \begin{bmatrix} K_F & J_R & J_R & J_R & J_R \\ a & K_I & J_G & 0 & 0 \\ b & e & K_S & J_G & 0 \\ c & 0 & f & K_T & J_G \\ d & 0 & 0 & g & K_E \end{bmatrix}$$

5.4.4 Calendrical Subsystem

The calendrical subsystem is responsible for displaying the cyclic records of both astronomical phenomena and the human festivities on the upper back dials of Antikythera device. The subsystem functions the 19-year Metonic cycle, the 76-year Callippic cycle, and the 4-year Olympiad cycle [8, 9]. The excavated structure relative to the calendrical subsystem is incomplete. The reconstructed mechanism remains unclear until the decoding of the back upper dials. Figure 5.12 shows the schematic drawing of the calendrical subsystem. The characteristics of the topological structure of the calendrical subsystem are concluded as follows:

1. It is a planar 6-bar mechanism with 9 joints and 1 degree of freedom.
2. It has a ground link (member 1, K_F), an input link (member 2, K_I), a Metonic cycle link (member 3, K_M), an Olympiad cycle link (member 4, K_O), a Callippic cycle link (member 5, K_C) and a transmission link (member 6, K_T). The Metonic cycle link, the Olympiad cycle link, and the transmission link are driven links. The Metonic cycle link and the Olympiad cycle link, respectively, show the Metonic cycle and the Olympiad cycle. And the Callippic cycle link is the output link to show the Callippic cycle.
3. It has 5 revolute joints (joints *a*, *b*, *c*, *d*, and *e*; J_R) and 4 gear joints (joints *f*, *g*, *h*, and *i*; J_G).
4. It is an ordinary gear train formed exclusively by external gears.

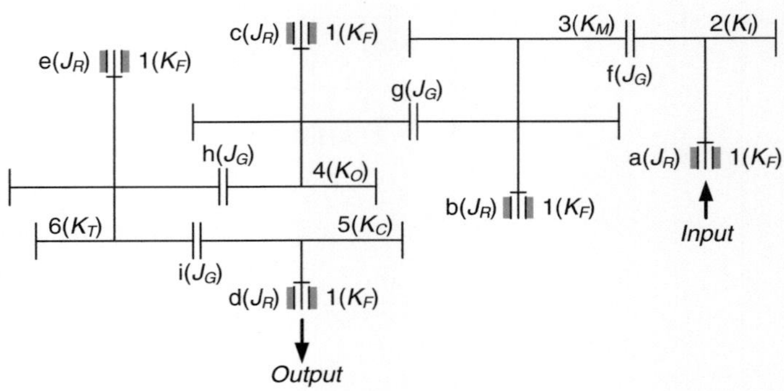

Fig. 5.12 Schematic drawing of the calendrical subsystem

5. The Metonic cycle link rotates once in every 19/5 years, the Olympiad cycle link rotates once in 4 years, and the Callippic cycle link rotates once in every 76 years.

The topology matrix of the calendrical subsystem (M_T) is:

$$M_{\mathrm{T}} = \begin{bmatrix} K_F & J_{\mathrm{R}} & J_{\mathrm{R}} & J_{\mathrm{R}} & J_{\mathrm{R}} & J_{\mathrm{R}} \\ a & K_{\mathrm{I}} & J_{\mathrm{G}} & 0 & 0 & 0 \\ b & f & K_{\mathrm{M}} & J_{\mathrm{G}} & 0 & J_{\mathrm{G}} \\ c & 0 & g & K_{\mathrm{O}} & 0 & 0 \\ d & 0 & 0 & 0 & K_{\mathrm{C}} & J_{\mathrm{G}} \\ e & 0 & h & 0 & i & K_{\mathrm{T}} \end{bmatrix}$$

The design of calendrical subsystem is the same as the existing design by Freeth et al. Nevertheless, as mentioned above, the surviving fragments corresponding to the calendrical subsystem are incomplete. Only two gears, respectively, with 15 and 60 teeth are identified form the surviving objects. Even though the above design (the 6-bar mechanism with 9 joints) is reasonable and satisfied the decoded functions, it is obvious that the surviving evidence is not sufficient enough to support the design completely.

Focusing on the uncertain situation of the calendrical subsystem, the related reconstruction work has to be investigated further. This subsystem with incomplete surviving evidence is introduced in Chap. 6. The possible topological structures, including the numbers of members and joints, have to be discussed according to the analysis of modern mechanisms and the identifications of gear ratios. Thus, all feasible reconstruction designs could be systematically generated through the design procedure introduced in Sect. 5.2. And, the tooth calculation for each obtained design can be obtained accordingly.

5.4.5 Lunar Subsystem

The lunar subsystem is to simulate the cyclic motion of the Moon on the front zodiac dial. The resulting motion is able to simulate lunar anomaly motion in accordance with Hipparchus's first lunar theory [8, 10]. With respect to the reconstruction design shown in Fig. 3.9, this subsystem locates between the fifth plate and the sixth plate (back plate) and connects the calendrical subsystem and the eclipse prediction subsystem. In this subsystem, there are 6 confirmed gears, respectively, with 223, 188, 50, 50, 50, and 50 teeth and 1 hypothetical gear with 27 teeth. The surviving gear with 188 teeth is just like a ring and mechanically connects to the other surviving gear with 223 teeth to form a compound gear. The shafts of the lunar subsystem have a complicated coaxial structure.

The reconstruction design of the lunar subsystem here is the existing reconstruction design by Freeth et al. [8]. Figure 5.13 shows its schematic drawing of

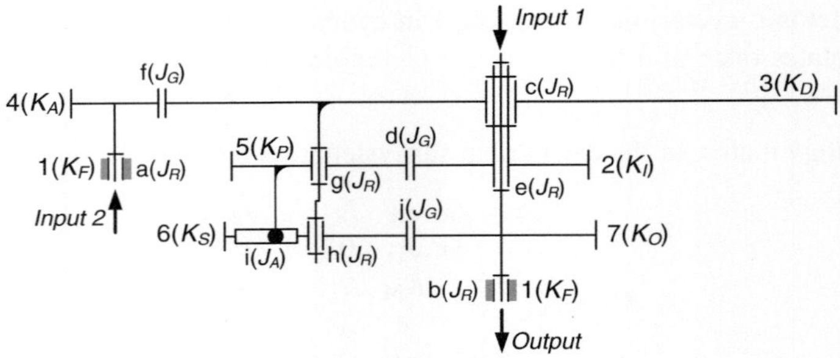

Fig. 5.13 Schematic drawing of the lunar subsystem

the lunar subsystem. Based on the analysis of the topological structure, the characteristics of topological structure are concluded as follows:

1. It is a planar 7-bar mechanism with 10 joints and 2 degrees of freedom.
2. It consists of a ground link (member 1, K_F), a sidereal gear, (member 2, K_I), a deferent gear (member 3, K_D), an anomalistic gear (member 4, K_A), an output gear (member 7, K_O), and a pin-in-slot device (K_{PS}) with a pin gear (member 5, K_P) and a slot gear (member 6, K_S). The sidereal gear and the anomalistic gear are the input links, and the output gear is the output link to show the lunar motion.
3. It has 9 joints, including 6 revolute joints (joints *a*, *b*, *c*, *e*, *g*, and *h*; J_R), 3 gear joints (joints *d*, *f*, and *j*; J_G), and 1 cam joint, i.e., a pin-in-slot joint (joint *I*; J_A).
4. It is a reverted epicyclic gear train, i.e., the input and output members are coaxial, and all gears are external gears.
5. The sidereal gear, which connects to the existing gear *e*2 and rotates once in a sidereal month, and the anomalistic gear, which connects to the hypothetical gear *m*3 and rotates once in a anomalistic month, are 2 input links.

The topology matrix of the lunar subsystem (M_T) is:

$$M_T = \begin{bmatrix} K_F & 0 & 0 & J_R & 0 & 0 & J_R \\ 0 & K_I & J_R & 0 & J_G & 0 & J_R \\ 0 & c & K_D & J_G & J_R & J_R & 0 \\ a & 0 & f & K_A & 0 & 0 & 0 \\ 0 & d & g & 0 & K_P & J_A & 0 \\ 0 & 0 & h & 0 & i & K_S & J_G \\ b & e & 0 & 0 & 0 & j & K_O \end{bmatrix}$$

The pin-in-slot joint, with two motion directions, is an ingenious mechanical design. A pin running through a slot allows the 2 joined mechanical members to revolute around each other and to translate with respect to each other in one direction. Based on the kinematic analysis of the pin-in-slot device, it is proved that the output of the pin-in-slot device successively introduces the motion with variable angular speed through the variation of rotational radius *r*. The variation period of the pin-in-slot device is related to its common frame.

Based on the above kinematic characteristics of the pin-in-slot device, the finding of such a joint from the excavation is a key factor to confirm the lunar subsystem with the characteristic of variable angular speed. As described in Chap. 1, Hipparchus developed two equivalent lunar theories according to the concept that the true lunar motion exhibits a simple periodic anomaly. Indeed, the mechanism of the above lunar subsystem completely demonstrates the lunar theory. Its output is the lunar motion with the rate of the sidereal month plus the harmonic variation of the anomalistic month. By engaging with the anomalistic gear (member 4, K_A), the deferent gear (member 3, K_D), i.e., the carrier of the epicyclic gear train, rotates with a rate difference between the anomalistic month and the sidereal month, i.e., the rate of the Moon's apogee. Furthermore, it is necessary to mount two gears incident to the pin-in-slot joint on the deferent gear. Thus, the period of variable rate (resulting by the variable rotation radius of the link with slot) would change from the period of sidereal month, which would occur when two gears incident to the pin-in-slot joint are on the fixed axes, to the anomalistic month, according to the two known inputs.

Even though the corresponding teeth of the mechanism are almost completely confirmed for the lunar subsystem, there is still an unexplained feature finding on the picture of surviving evidence [8]. The feature is a visible regular pentagon in the center of a gear, as shown in Fig. 5.14. From the design viewpoint of mechanical elements, this unexplained feature would affect the identifications of links in the lunar subsystem. In addition, the arrangement of multiple coaxial axes in the design strongly relates to the manufacturing possibility in ancient time.

Therefore, the above crucial reasons make Yan and Lin reconstruct all feasible designs of the lunar subsystem based on the systematic reconstruction design procedure described in Sect. 5.2. Furthermore, the possible resolutions for the unexplained feature would be proposed in accordance with the structure analysis of the obtained feasible reconstruction designs. The detail reconstruction work is described in Chap. 7.

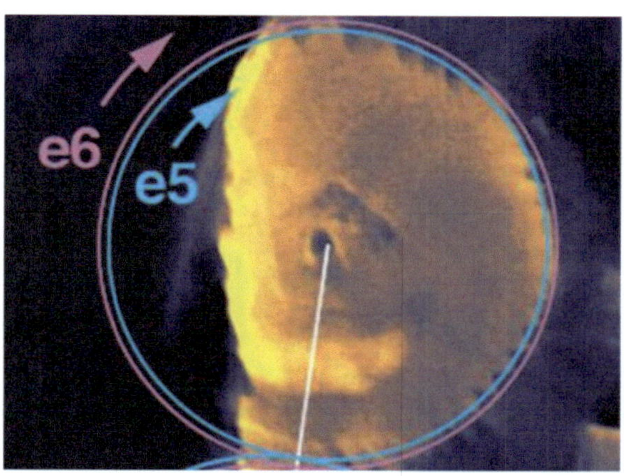

Fig. 5.14 An unexplained feature in the surviving fragment. Reprinted from Ref. [8], copyright 2015, with permission from Nature

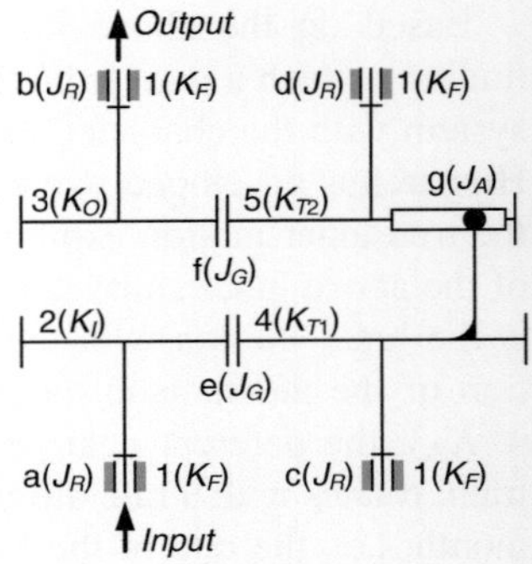

Fig. 5.15 Schematic drawing of the solar subsystem

5.4.6 Solar Subsystem

The solar subsystem is a part of the lost structures of the excavation. As the meaning of its name, this subsystem is defined to demonstrate the solar anomaly motion in accordance with the appropriate astronomical theory in the era of Antikythera device. Figure 5.15 shows the schematic drawing of the solar subsystem in a complete reconstruction design shown in Fig. 3.9. Based on the existing design, the characteristics of the topological structure of the solar mechanism are as follows:

1. It is a planar 5-bar mechanism with 7 joints and 1 degree of freedom.
2. It has a ground link (member 1, K_F), an input link (member 2, K_I), an output link (member 3, K_O), and 2 transmission links (members 4 and 5, K_{T1} and K_{T2}).
3. It has 4 revolute joints (joints *a*, *b*, *c*, and *d*; J_R), 2 gear joints (joints *e* and *f*; J_G), and 1 pin-in-slot joint (joint *g*; J_A).
4. It is an ordinary gear train formed exclusively by external gears.
5. The input link has to rotate clockwise in the period of a tropical year.

The topology matrix of the solar subsystem (M_T) is:

$$M_T = \begin{bmatrix} K_F & J_R & J_R & J_R & J_R \\ a & K_I & 0 & J_G & 0 \\ b & 0 & K_O & 0 & J_G \\ c & e & 0 & K_{T1} & J_A \\ d & 0 & f & g & K_{T2} \end{bmatrix}$$

5.4.7 Planetary Subsystem

The planetary subsystem is also a part of the lost structures of the excavation. This subsystem is more complicated than the solar subsystem. It has to include several mechanisms, respectively, to generate the anomaly motions of Mercury, Venus, Mars, Jupiter, and Saturn with special periods. These mechanisms are designated based on the Apollonius' epicyclic system mentioned in Chap. 1. Thus, for the defined planetary subsystem, its resulting motions are completely the planetary motions of the Apollonius' epicyclic system.

Based on the introduction of the ancient astronomy in Chap. 1, it is known that the inferior planets and the superior planets have respective kinematic characteristics by the common Apollonius' epicyclic system. Therefore, in the planetary subsystem of the complete reconstruction design shown in Fig. 3.9, two different reconstruction designs are used to demonstrate the motions of the inferior planets and the superior planets, respectively. Figure 5.16a shows the schematic drawing of a reconstruction design for the inferior planets, and Fig. 5.16b shows the mechanical drawing of a reconstruction design for the superior planets.

The topological characteristics of the design of the inferior planets shown in Fig. 5.16a are concluded as follows:

1. It is a planar 4-bar mechanism with 5 joints and 1 degree of freedom.
2. It has a ground link (member 1, K_F), an input link (member 2, K_I), an output link (member 3, K_O), and a transmission link (member 4, K_T).
3. It has 3 revolute joints (joints *a*, *b*, and *d*; J_R), 1 gear joint (joint *c*, J_G), and 1 pin-in-slot joint (joint *e*, J_A).
4. It is an epicyclic gear train and formed exclusively by external gears.
5. The ground link is a sun gear.
6. The input link is the carrier of the epicyclic gear train and rotates once in a year.

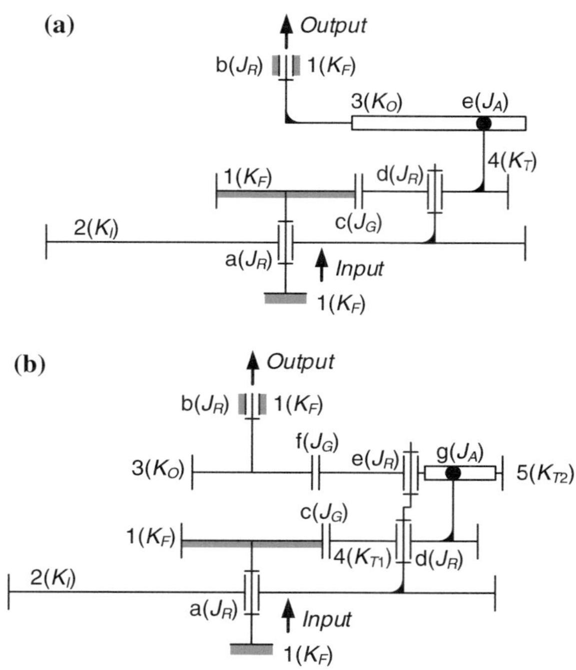

Fig. 5.16 Schematic drawing of the planetary subsystem. **a** Inferior planets. **b** Superior planets

The topology matrix of the inferior planetary mechanisms (M_T) is:

$$M_T = \begin{bmatrix} K_F & J_R & J_R & J_G \\ a & K_I & 0 & J_R \\ b & 0 & K_O & J_A \\ c & d & e & K_T \end{bmatrix}$$

The topological characteristics of the design of the superior planets shown in Fig. 5.16b are concluded as follows:

1. It is a planar 5-bar mechanism with 7 joints and 1 degree of freedom.
2. It has a ground link (member 1, K_F), an input link (member 2, K_I), an output link (member 3, K_O), and 2 transmission links (members 4 and 5; K_{T1} and K_{T2}).
3. It has 4 revolute joints (joints *a*, *b*, *d*, and *e*; J_R), 2 gear joints (joints *c* and *f*, J_G), and 1 pin-in-slot joint (joint *g*, J_A).
4. It is an epicyclic gear train formed exclusively by external gears.
5. The ground link is the sun gear.
6. The input link is the carrier of the epicyclic gear train and rotates with the rates of the designated planets.

The topological matrix of the superior planetary mechanisms (M_T) is:

$$M_T = \begin{bmatrix} K_F & J_R & J_R & J_G & 0 \\ a & K_I & 0 & J_R & J_R \\ b & 0 & K_O & 0 & J_G \\ c & d & 0 & K_{T1} & J_A \\ 0 & e & f & g & K_{T2} \end{bmatrix}$$

Based on the above structure analysis of mechanisms of the Yan and Lin's reconstruction designs, respectively, for the solar motion, the inferior motion, and the superior motion, it is clear that the design concept of epicyclic gear trains with a pin-in-slot joint is a common concept. In addition, the pin-in-slot joint is indispensable to generate anomaly motions. However, the truth is that the corresponding evidence is completely lost. Even though the above reconstruction designs for the lost subsystems are reasonable and satisfy the standard of science and technology in the era of Antikythera device, they do not confirm absolutely as the original designs. Due to the losing of corresponding evidence, it is believed that these lost subsystems can be achieved by different design concepts. From the above reasons, the systematic reconstruction of all feasible designs of the solar subsystem and the planetary subsystem is presented in Chaps. 8 and 9, respectively.

5.4.8 Summary

In summary, all subsystems and their existing problems are presented through introducing the topological analysis of the interior structure of Yan and Lin's reconstruction designs. It is understood that the calendrical subsystem is incomplete, the lunar subsystem has an unexplained feature subsystem, and the solar and the planetary subsystems are lost. For these questionable subsystems, their design specifications and constraints are concluded through the detected evidence, the decoded information, ancient astronomy, ancient astronomical instruments, and kinematic and mechanism analyses. Then, in accordance with the concepts of generalization and specialization, all feasible topological structures subject to the design constraints, for the unclear and lost mechanisms, can be synthesized. Certainly, new reconstruction designs with complete interior mechanisms are generated by combining the feasible design concepts of the unclear and lost mechanisms and the existing mechanisms.

These questionable subsystems resulting in the corresponding reconstruction works are presented in Chaps. 6–9. Furthermore, the assembly work for all possible reconstruction designs of the mechanism is presented in Chap. 11.

5.5 Comparisons Among Different Reconstruction Researches

Here, the reconstruction shown in Fig. 3.9 by Yan and Lin is used to compare with several existing reconstruction designs in different time stages. For the same functions, the mechanisms applied in these reconstruction designs are introduced. And for the six subsystems, the contributions of these reconstruction designs are concluded as listed in Table 5.1.

Table 5.1 Comparisons among existing reconstruction designs

		Diverse reconstruction designs				
		Price	Edmund and Morgan	Wright	Freeth et al.	Yan and Lin
Interior subsystems	Date	○	○	◎	◎	◎
	Eclipse	△	△	○	◎	◎
	Lunar	○	○	○	◎	◎
	Calendrical	△	△	△	✓	✓
	Solar	○	○	✓	✓	✓
	Planetary	●	△	✓	✓	✓

◎ Reconstructed; confirmed by surviving evidence
✓ Reconstructed; lost surviving evidence or incomplete surviving evidence
○ Reconstructed; incorrect design, identification, or position
△ Incompletely reconstructed; no identification, incorrect position, or identification
● Unreconstructed

5.5.1 Comparison with Price's Design

There are several differences between Price's reconstruction design and Yan and Lin's reconstruction design shown in Fig. 3.9. For the date subsystem, the corresponding gear mechanism of Yan and Lin's design appears in Price's design. However, Price's design does not mention the existing of the date pointer for indicating the date on the front Egyptian calendar dial. This gear mechanism is only used to drive the gear train by engaging the contrate gear. For the eclipse prediction subsystem, whatever Yan and Lin's design or Price's design, the corresponding gear mechanisms are the same, i.e., an ordinary gear train with 5 links, 4 revolute joints, and 3 gear joints. Some gears contained in this subsystem are identified to have the same teeth. However, the purposes of these two reconstruction designs are obviously different.

For the lunar subsystem, the gear mechanism in Yan and Lin's design is completely different from the gear mechanism to display the lunar motion in Price's design. Except for the dissimilar structure, the gear mechanism in Price's design cannot generate the lunar motion with the characteristic of anomaly. Moreover, none of the available historical literatures prove the application of differential gear trains in the early historical development of mechanisms, even though such design is a well-known mechanical design in modern times. For this reason, this viewpoint seems to be arguable. For the calendrical subsystem, owing to the unknown purposes of the upper back dials, its corresponding gear mechanism is incomplete in Price's design.

For the solar subsystem, both two reconstruction designs can display the solar motion, but they use different gear mechanisms to imitate such a function. The most important is that the solar motion in Price's design is a cyclic motion with a mean rate, which is not able to show an anomaly motion. Finally, for the planetary subsystem, the gear mechanisms for the planetary motions are inexistent, and Price did not present such functions.

5.5.2 Comparison with Edmund and Morgan's Design

It is understood that Edmund and Morgan's reconstruction design shown in Fig. 4.2 is on the basis of Price's design. Certainly, the differences between the Price's design and the Yan and Lin's design existed in the Edmund and Morgan's design. Edmund and Morgan proposed a further design for the motions of Venus and Mars. The corresponding gear mechanisms should belong to the planetary subsystem, and they are located at different positions of the interior mechanisms.

Based on the concept of modern mechanism, the gear mechanisms for Venus and Mars are the same in both Edmund and Morgan's design and Yan and Lin's design. They are epicyclic gear trains with a pin-in-slot joint. Nevertheless, there is a difference in these two designs according to the position of interior mechanisms.

Edmund and Morgan rebuilt the gear mechanism for Mars in the position of the calendrical subsystem, since they did not realize the purpose of the upper back dials at that time.

5.5.3 Comparison with Wright's Design

In Sect. 4.4, Wright's reconstruction work was introduced and compared with Price's existing design. Here, the comparison of the six interior subsystems between Wright's reconstruction design and the Yan and Lin's reconstruction design is presented.

The reconstructed mechanisms of the date subsystem and the eclipse prediction subsystem are identical in these two designs. The gear mechanism of the lunar subsystem is misunderstood in Wright's design. The mass of gears located at the lunar subsystem of Yan and Lin's design, i.e., it is the lunar mechanism presented by Freeth et al. Firstly, it is identified as an epicyclic gear train. However, its purpose seems to provide the power for the eclipse prediction subsystem of the lower dials, not to generate the lunar motion. The pin-in-slot joint, the key inside this mass of gears, was discovered in Wright's work, but an appropriate explanation of its application was not proposed. In addition, an idle planet gear exists in the epicyclic gear train of Wright's design. Therefore, the reconstruction work of the lunar subsystem between these two designs is different. As to the calendrical subsystem, Wright knew that there is an ordinary gar train, like his reconstruction design of the Byzantine sundial with calendrical device, for calculating some astronomical cycles. However, this design did not contain the Olympiad cycle.

Since Wright assumed that the gear mechanisms of the planetary subsystem and the solar subsystem should have existed in the Antikythera device, the corresponding reconstruction mechanisms appear in Wright's design. For the planetary subsystem, the mechanisms in Wright's design for Mars, Jupiter, and Saturn are different to those of Yan and Lin's design. Wright used more gears to exchange the speed rates so as to generate the necessary output rates. But, the mechanisms for Venus and Mercury are the same in both reconstruction designs. Furthermore, the gear mechanisms of the solar subsystem are unlike in these two reconstruction designs. Wright used an epicyclic gear train with a pin-in-slot joint for the solar motion, but Yan and Lin used an ordinary gear train with a pin-in-slot joint and a stepped arbor for such a function. In fact, focusing on the solar subsystem, Yan and Lin presented several possible reconstruction designs based on their systematic design methodology. And these obtained designs include Wright's design. The detail presentation regarding the reconstruction work of the solar subsystem is in Chap. 8. In a word, the mechanisms of the solar subsystem, respectively, shown in Wright's design and Yan and Lin's design are different, as shown in Fig. 3.9. The mechanism of the solar subsystem in the Wright's design is the same as the solar subsystem in Yan and Lin's reconstruction designs shown in Appendix A.

5.5.4 Comparison with the Design of Freeth et al.

The achievements of reconstruction research by Freeth et al. (AMRP) are the essential basis of Yan and Lin's reconstruction research. According to the decoded exterior functions, defined appearance, and reconstructed interior mechanism of the design by Freeth et al., Yan and Lin established most of the historical archives in the first stage of their systematic reconstruction design methodology. Furthermore, they conclude the six interior subsystems by the concept of mechanisms. Therefore, the reconstruction design of Freeth et al. as shown in Fig. 4.9 and the reconstruction designs by Yan and Lin have several common designs regarding the interior structure of mechanisms.

The date subsystem, the eclipse prediction subsystem, the lunar subsystem, and the calendrical subsystem exist in the reconstruction design of Freeth et al. The mechanisms of the date subsystem, the eclipse prediction subsystem, and the lunar subsystem are the same in both the Freeth et al.'s reconstruction design and that of Yan and Lin. But, focusing on the lunar subsystem with unexplained feature, Yan and Lin present some possible explanations for this feature in Chap. 7.

Furthermore, the calendrical subsystem in these two reconstruction designs has different reconstructed mechanisms. The calendrical subsystem is with incomplete surviving evidence. Even though its reconstructed mechanism in the design of Freeth et al. is reasonable, it is not confirmed absolutely that this subsystem has such a reconstruction design. Therefore, Yan and Lin present the other possible result by considering the same numbers of gears. In their designs, the subsidiary dial for the Callippic cycle can have opposite display direction, i.e., this subsidiary dial in the Freeth et al's design and Yan and Lin's design displays counterclockwise and clockwise, respectively. In fact, Yan and Lin simultaneously obtained these two possible designs of the calendrical subsystem by their design methodology that is represented in Chap. 6. Regarding the solar and the planetary subsystems, Freeth et al's design shown in Fig. 4.9 does not present the corresponding reconstructed mechanisms. But, they indicate that the interior structure could have lost epicyclic gear trains for demonstrating these functions. These lost epicyclic gear trains are designed based on Hipparchus astronomical theories.

As a matter of fact, for the solar and the planetary subsystems, Freeth and Jones did not present a partial reconstruction design until 2012 [11]. In Freeth and Jone's design, as well as in Yan and Lin's design, the reconstructed gear mechanisms of the planetary subsystem, respectively, for the inferior planets and the superior planets have the same topological structure. The differences are the space arrangement of the interior structure and the reconstructed teeth. As to the solar subsystem, these two reconstruction designs are different. Freeth and Jone apply a design concept of epicyclic gear trains, but Yan and Lin apply a design concept of ordinary gear trains. The reconstruction design of the solar subsystem by Freeth and Jones is also included in the results of Yan and Lin's systematic reconstruction research in Chap. 8.

5.6 Remarks

The reconstruction research of ancient mechanisms contains the study of historical archives and reconstruction designs which includes reconstruction analysis and reconstruction synthesis. This process is contributed to define the problems of reconstruction research of ancient mechanisms, to develop appropriate design specifications and design constraints, to provide the analytical results for defining the ancient technology by modern knowledge, methods and equipment, and to generate all possible designs subject to the science and technology at that time.

A systematic reconstruction design methodology of lost ancient mechanisms containing the build of historical archives and four design steps is presented. This methodology can be used for the reconstruction design of the interior mechanisms of Antikythera device. The historical archives of Antikythera device include detected evidence, decoded information, ancient astronomy, ancient astronomical instruments, and kinematic and mechanism analyses. All concluded information is useful for each design step of the design methodology.

In summary, the reconstruction research of the Antikythera device by Yan and Lin is introduced. The structure analysis of mechanisms of the reconstruction designs with complete mechanisms resulting from the design methodology is presented. And the surviving evidence corresponding to the six interior subsystems is described to define the problems, i.e., the subsystems are confirmed, unclear, incomplete, or lost subsystems. Finally, the comparisons among the reconstructed designs by Yan and Lin and other existing designs are discussed.

References

1. Yan HS (2007) Reconstruction designs of lost ancient Chinese machinery. Springer, Netherland
2. Yan HS (1992) A methodology for creative mechanism design. Mech Mach Theory 27(3):235–242
3. Yan HS (1998) Creative design of mechanical devices. Springer, Singapore
4. Lin JL (2011) Systematic reconstruction designs of Antikythera mechanism. Dissertation, National Cheng Kung University, Tainan, Taiwan
5. Yan HS, Wu LI (2014) Mechanisms (in Chinese), 4th edn. Dong Hua Books, Taipei
6. Cleghorn WL (2005) Mechanics of machines. Oxford University Press, New York
7. de Solla Price D (1974) Gears from the Greeks: the Antikythera mechanism—a calendar computer from ca. 80 BC. Science History Publications, New York
8. Freeth T et al (2006) Decoding the ancient Greek astronomical calculator known as the Antikythera mechanism. Nature 444:587–891
9. Freeth T et al (2008) Calendars with olympiad display and eclipse prediction on the Antikythera mechanism. Nature 454:614–617
10. Freeth et al T (2006) Decoding the ancient Greek astronomical calculator known as the Antikythera mechanism. Nature 444:587–891. (Supplementary Notes)
11. Freeth T, Jones A (2012) The cosmos in the Antikythera mechanism. In: Institute for the Study of the Ancient World, New York University. Available in the website: Institute for the Study of the Ancient World (ISAW). http://dlib.nyu.edu/awdl/isaw/isaw-papers/4/. Accessed Jul 2014

Chapter 6
Reconstruction Designs of the Calendrical Subsystem

Although a reasonable design of the calendrical subsystem by Freeth et al. [1–3] was presented, the weak surviving evidence cannot support this design completely. Indeed, only two gears, respectively, with 15 and 60 teeth are identified from the excavation. Focusing on the calendrical subsystem with incomplete surviving evidence, this chapter applies the design methodology presented in Chap. 5 to generate all feasible reconstruction designs of the calendrical subsystem [4–7].

First, the historical archives about the calendrical subsystem are introduced. The mechanism and kinematic analyses are contributed to conclude the possible topological structures of mechanisms. The confirmed astronomical cycles included in the calendrical subsystem are the known member rates of the subsystem. By the analysis of gear ratios, the relative design constraints are concluded. And, the reconstruction designs of calendrical subsystem subject to the standard and technique and science at the time of the Antikythera device are obtained.

6.1 Historical Archives of the Calendrical Subsystem

Before illustrating the reconstruction design procedure of the calendrical subsystem, the related research to build the corresponding historical archives is prepared. The functions, the surviving evidence, and the astronomical cycles contained in the subsystem are understood. Also, ancient astronomical instruments with similar functions are studied. According to the analysis of the existing designs, the calendrical subsystem is known as an ordinary gear train. However, the number of mechanical elements to construct the calendrical subsystem is uncertain. Therefore, in order to define appropriate design specifications of the calendrical subsystem so as to demonstrate the design procedure, the possible numbers of members and joints must be discussed first, especially for the mechanism analysis of ordinary gear trains.

J.-L. Lin and H.-S. Yan, *Decoding the Mechanisms of Antikythera Astronomical Device*, DOI 10.1007/978-3-662-48447-0_6

Based on mobility analysis, the following expression is true for a planar calendrical mechanism with 1 degree of freedom, N_{L} members, N_{JR} revolute joints, and N_{JG} gear joints:

$$2N_{\mathrm{JR}} + N_{\mathrm{JG}} - 3N_{\mathrm{L}} + 4 = 0 \tag{6.1}$$

And, it is obvious that the following expression is true for a gear train with N_{J} joints:

$$N_{\mathrm{JR}} + N_{\mathrm{JG}} - N_{\mathrm{J}} = 0 \tag{6.2}$$

Since every gear in a gear train must at least have a revolute joint, the following expression is true:

$$N_{\mathrm{JR}} - N_{\mathrm{L}} + 1 = 0 \tag{6.3}$$

By solving Eqs. (6.1), (6.2), and (6.3) in terms of the number of links, the following relations for the numbers of joints are concluded:

$$N_{\mathrm{J}} = 2N_{\mathrm{L}} - 3 \tag{6.4}$$

$$N_{\mathrm{JR}} = N_{\mathrm{L}} - 1 \tag{6.5}$$

$$N_{\mathrm{JG}} = N_{\mathrm{L}} - 2 \tag{6.6}$$

Since the calendrical subsystem includes three dials to display different periods, the number of gear joints in the reconstruction design of the mechanism is 3 at least in order to achieve the transformations of rotation rates. Thus, the number of links in the calendrical subsystem of Antikythera device should be at least 5, based on Eq. (6.6). From designers' viewpoint, a simpler design is a better design for mechanisms with the same output functions. Furthermore, two design concepts with simpler structure are concluded. One is the mechanisms with 5 members, 7 joints (consisting of 4 revolute joints and 3 gear joints), and 1 degree of freedom. The other is the mechanisms with 6 members, 9 joints (consisting of 5 revolute joints and 4 gear joints), and 1 degree of freedom.

Gear trains are used to transmit torque and rotary motion with specific ratios. For meshing gears, the gear ratio (speed ratio) is inversely proportional to the teeth ratio. It is well-known that the rate of some links of the calendrical subsystem must satisfy the required cycles. Therefore, once the rates of meshing gears are confirmed, the corresponding teeth relation can be concluded. In the following, the kinematic analysis of gear trains is carried out to conclude the teeth relations between different links.

The confirmed interior mechanism indicates that the input link of the calendrical subsystem rotates counterclockwise at $53/(19 \times 3)$ rotations per year. Based on the decoding of the inscriptions on the exterior dials, some rotation rates and directions of links have been confirmed [1, 3]. The Metonic cycle link rotates clockwise at 5/19 rotations per year. The Olympiad cycle link rotates counterclockwise at 1/4 rotations per year. Furthermore, the meanings of the inscriptions on the Callippic cycle dial are still unknown. The only confirmed fact is that the

Callippic cycle link must rotate at 1/76 rotations per year without defining its rotation direction. Hence, in accordance with these defined rates and the direction of link, the engagement conditions of gears are discussed to conclude the possible topological characteristics and the appropriate design constraints. By the surviving evidence and the known conditions, six cases of different meshing gears are analyzed as follows:

1. When a gear joint is incident to the input link (member 2) and the Metonic cycle link (member 3), the following expression is true for its gear ratio:

$$[53/(19 \times 3)] \times N_2/N_3 = 5/19$$

and

$$N_2/N_3 = 15/53$$

2. When a gear joint is incident to the input link (member 2) and the Olympiad cycle link (member 4), the following expression is true for its gear ratio:

$$[53/(19 \times 3)] \times N_2/N_4 = 1/4$$

and

$$N_2\big/N_4 = 57\big/212$$

3. When a gear joint is incident to the input link (member 2) and the Callippic cycle link (member 5), the following expression is true for its gear ratio:

$$[53/(19 \times 3)] \times N_2/N_5 = 1/76$$

and

$$N_2/N_5 = 3/212$$

4. When a gear joint is incident to the Metonic cycle link (member 3) and the Olympiad cycle link (member 4), the following expression is true for its gear ratio:

$$5/19 \times N_3/N_4 = 1/4$$

and

$$N_3/N_4 = 19/20$$

5. When a gear joint is incident to the Metonic cycle link (member 3) and the Callippic cycle link (member 5), the following expression is true for its gear ratio:

$$5/19 \times N_3/N_5 = 1/76$$

and

$$N_3/N_5 = 1/20$$

6. When a gear joint is incident to the Olympiad cycle link (member 4) and the Callippic cycle link (member 5), the following expression is true for its gear ratio:

$$1/4 \times N_4/N_5 = 1/76$$

and

$$N_4/N_5 = 1/19$$

Different gear ratios result in differences in teeth number and gear size between two meshing gears. It is suggested that the Callippic cycle link should not be adjacent to the input link, the Metonic cycle link or the Olympiad cycle link, to avoid generating too big a gear. Therefore, there should be a link adjacent to the Callippic cycle link, i.e., the simplest topological structure of the calendrical subsystem should be a planar mechanism with 1 degree of freedom, 6 links, and 9 joints, under the particular constraints of geometrical sizes.

6.2 Design Process of the Calendrical Subsystem

In this section, the design procedure of the calendrical subsystem is illustrated step by step. The related design specifications and design constraints are based on solid previous studies. Certainly, all feasible reconstructions are obtained through this design procedure. Moreover, the tooth calculation relations for each reconstructed gear train are also derived.

6.2.1 Design Specifications

In the stage of building historical archives, it is concluded that the minimum number of members of the calendrical subsystem is six through the mechanism and kinematic analyses. The concept of the research is to adopt the simplest structure to reconstruct the incomplete subsystem. Here, the design specifications of the calendrical subsystem, which satisfy these confirmed functions, are concluded as follows:

1. It is a planar 6-bar mechanism with 9 joints and 1 degree of freedom.
2. The mechanical members are gears and links.
3. The joints consist of 5 revolute joints and 4 external gear joints.

6.2.2 Generalized Chains

Figure 6.1 shows the existing design by Freeth et al. and its generalized chain with 6 links and 9 joints. The ground link (member 1) is generalized into a quinary link, the input link (member 2) is generalized into a binary link, the Metonic cycle link (member 3) is generalized into a quaternary link, the Olympiad period link (member 4) is generalized into a binary link, the Callippic period link (member 5) is generalized into a binary link, and the transmission link (member 6) is generalized into a ternary link.

Based on the concept of number synthesis, the atlas of generalized chains with 6 links and 9 joints can be generated. According to the concluded design constraints of the ground link, each link of the calendrical subsystem must be adjacent

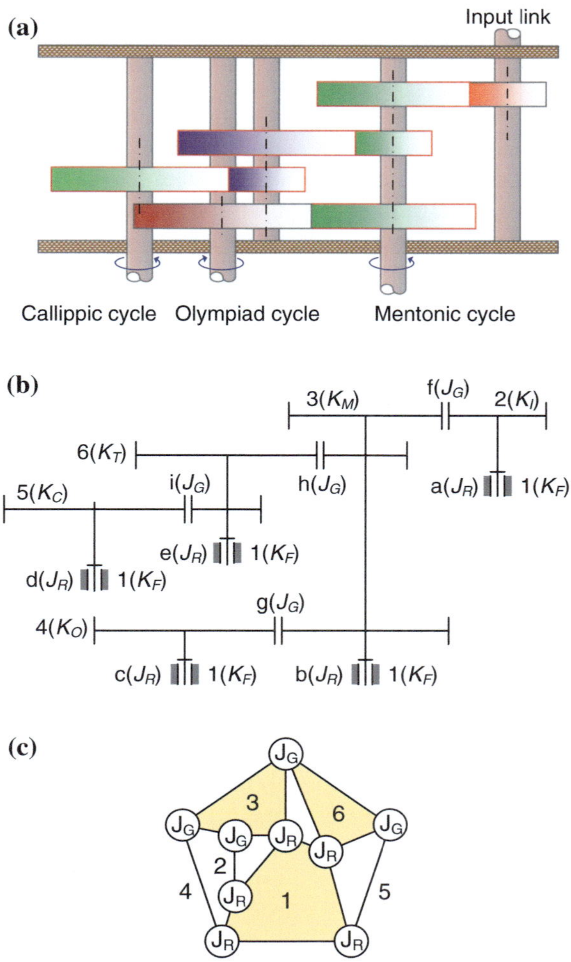

Fig. 6.1 Calendrical subsystem reconstructed by Freeth et al. **a** Existing design. **b** Schematic drawing. **c** Generalized chain

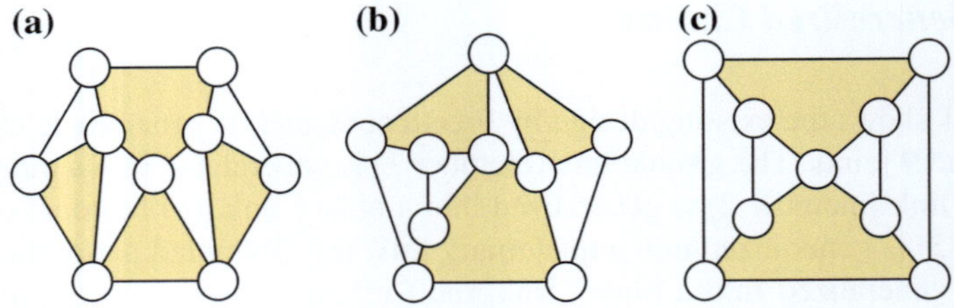

Fig. 6.2 Atlas of feasible generalized chains with 6 links and 9 joints

to the ground link with a revolute joint. Therefore, only 3 generalized chains with 6 links and 9 joints are feasible, as shown in Fig. 6.2.

6.2.3 *Specialized Chains*

In order to assign a specific link or joint into an available generalized chain, the corresponding design constraints have to be established first. It is known that each dial contained in the calendrical subsystem has its designated rotation rate. The above discussion regarding the gear ratios contributes to analyze whether the links are adjacent to each other with a gear joint. Therefore, according to the previous studies, the design constraints of the calendrical subsystem are concluded as follows:

1. The mechanism of the calendrical subsystem must demonstrate the Metonic cycle, the Olympiad cycle, and the Callippic cycle.
2. The mechanism must at least include the ground link, the input link, the Metonic cycle link, the Olympiad cycle link, the Callippic cycle link, and the transmission link, since it is defined as a planar 6-bar mechanism.

   Ground link (member 1, K_F)

   (a) There must be a ground link as the frame.
   (b) A ground link must be a multiple link in order to serve as the input and three confirmed functions displayed on the exterior dials.

   Input link (member 2, K_I)

   (a) There must be a link as the input link for the power transmission.
   (b) The input link must rotate counterclockwise at $53/(19 \times 3)$ rotations per year.
   (c) The input link must include a gear with 15 teeth. This is known by distinguishing the image of the surviving mechanism.
   (d) Previous analysis shows that the gear ratio for the input link to the Olympiad period link is 57/212 which conflicts with the existing evidence that the Olympiad cycle link must have a gear with 60 teeth. This means that the teeth number of the input link is not an integer. Therefore, the input link (link 2) must not be adjacent to the Olympiad cycle link (link 4).

(e) Based on the previous analysis of gear joints, the input link (link 2) must not be adjacent to the Callippic cycle link (link 5) with a gear joint, in order not to generate a large gear ratio.

Metonic cycle link (member 3, K_M)

(a) There must be a link as the Metonic cycle link.
(b) The Metonic cycle link rotates clockwise at 5/19 rotations per year.
(c) The Metonic cycle link is adjacent to the ground link with a revolute joint.
(d) The Metonic cycle link (link 3) must not be adjacent to the Callippic cycle link (link 5) with a gear joint, in order not to generate too large a gear ratio.

Olympiad cycle link (member 4, K_O)

(a) There must be a link as the Olympiad cycle link.
(b) The Olympiad cycle link rotates counterclockwise at 1/4 rotations per year.
(c) The Olympiad cycle link is adjacent to the ground link with a revolute joint.
(d) The Olympiad cycle link must have one gear with 60 teeth. This is understood by distinguishing the image of the surviving mechanism.
(e) The Olympiad cycle link (link 4) must not be adjacent to the Callippic cycle link (link 5) with a gear joint, in order not to generate too large a gear ratio.

Callippic cycle link (member 5, K_C)

(a) There must be a link as the Callippic cycle link.
(b) The Callippic cycle link rotates at 1/76 rotations per year.
(c) The Callippic cycle link is adjacent to the ground link with a revolute joint.

Transmission link (member 6, K_T)

(a) There must be a link as the transmission link in order to generate appropriate teeth numbers.
(b) The transmission link is adjacent to the ground link with a revolute joint.
(c) The transmission link must be at least a ternary link in order to avoid generating a redundant structure.
(d) When the transmission link is adjacent to both the input link and the Callippic cycle link, the following expression is true for its gear ratio:

$$[53/(19 \times 3)] \times N_2/N_5 \times N_5'/N_6 = 1/76$$

or

$$N_2/N_5 \times N_5'/N_6 = 3/212$$

where N_5 is the gear of the Callippic cycle link adjacent to the input link, and N_5' is the gear of the Callippic cycle link adjacent to the transmission link. This large gear ratio results in unsuitable gear sizes. Therefore, the transmission link (link 6) must not be adjacent to both the input link (link 2) and the Callippic cycle link (link 5).

Among the obtained 3 generalized chains, all possible specialized chains, subject to the concluded design constraints, are identified by the following steps:

Step 1. For each available generalized chain, the ground link and its incident revolute joints are assigned.
Step 2. For each result from Step 1, the Callippic cycle link and its incident gear joints are assigned.
Step 3. For each result from Step 2, the Olympiad cycle link and its incident gear joints are assigned.
Step 4. For each result from Step 3, the input link and its incident gear joint are assigned.
Step 5. For each result from Step 4, the Metonic cycle link and its incident joint are assigned.
Step 6. For each result from Step 4, the transmission link and its incident gear joints are assigned.

Thus, all possible specialized chains corresponding to the 3 feasible generalized chains with 6 links and 9 joints shown in Fig. 6.2 can be identified as follows.

6.2.3.1 Ground Link (Member 1)

Since there must be a multiple link with 5 revolute joints as the frame, the ground link is identified as follows:

1. For the generalized chain shown in Fig. 6.2a, the assignment of the ground link generates 1 result, Fig. 6.3a.
2. For the generalized chain shown in Fig. 6.2b, the assignment of the ground link generates 1 result, Fig. 6.3b.
3. For the generalized chain shown in Fig. 6.2c, the assignment of the ground link generates 1 result, Fig. 6.3c.

Therefore, 3 specialized chains with identified ground link are available, as shown in Fig. 6.3.

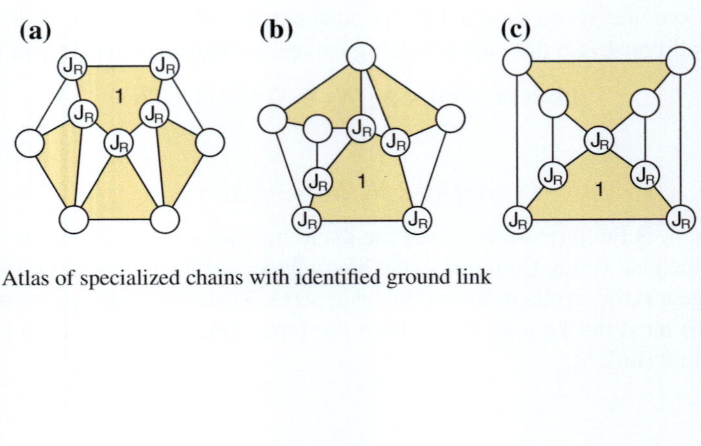

Fig. 6.3 Atlas of specialized chains with identified ground link

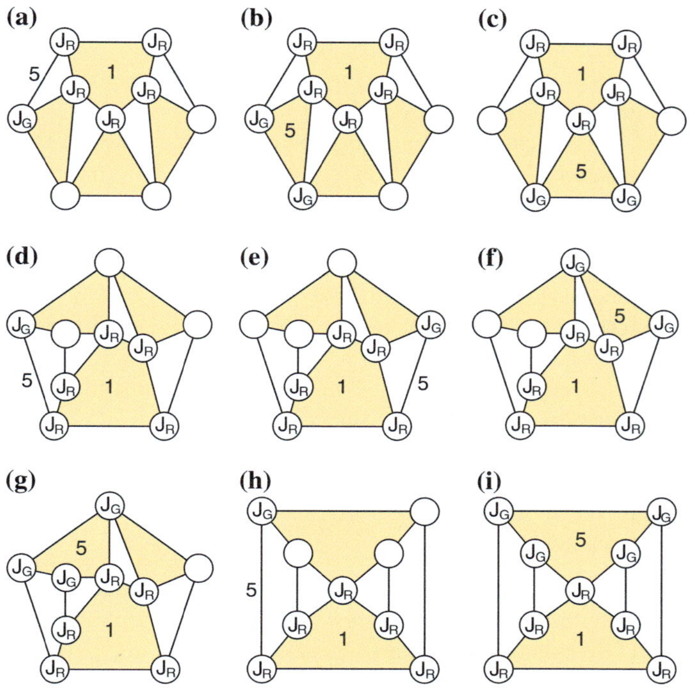

Fig. 6.4 Atlas of specialized chains with identified ground link and Callippic cycle link

6.2.3.2 Callippic Cycle Link (Member 5)

Since there must be a link as the Callippic cycle link, the Callippic cycle link is identified as follows:

1. For the generalized chain shown in Fig. 6.3a, the assignment of the Callippic cycle link generates 3 results, Fig. 6.4a–c.
2. For the generalized chain shown in Fig. 6.3b, the assignment of the Callippic cycle link generates 4 results, Fig. 6.4d–g.
3. For the generalized chain shown in Fig. 6.3c, the assignment of the Callippic cycle link generates 2 results, Fig. 6.4h–i.

Therefore, 9 specialized chains with identified ground link and Callippic cycle link are available, as shown in Fig. 6.4.

6.2.3.3 Olympiad Cycle Link (Member 4)

Since there must be a link as the Olympiad cycle link to display the Olympiad cycle, the Olympiad cycle link is identified as follows:

1. For the generalized chain shown in Fig. 6.4a, the assignment of the Olympiad cycle link generates 3 results, Fig. 6.5a–c.
2. For the generalized chain shown in Fig. 6.4b, the assignment of the Olympiad cycle link generates 2 results, Fig. 6.5d–e.
3. For the generalized chain shown in Fig. 6.4c, the assignment of the Olympiad cycle link generates 1 result, Fig. 6.5f.
4. For the generalized chain shown in Fig. 6.4d, the assignment of the Olympiad cycle link generates 3 results, Fig. 6.5g–i.
5. For the generalized chain shown in Fig. 6.4e, the assignment of the Olympiad cycle link generates 2 results, Fig. 6.5j–k.
6. For the generalized chain shown in Fig. 6.4f, the assignment of the Olympiad cycle link generates 1 result, Fig. 6.5l.
7. For the generalized chain shown in Fig. 6.4g, the assignment of the Olympiad cycle link generates 1 result, Fig. 6.5m.
8. For the generalized chain shown in Fig. 6.4h, the assignment of the Olympiad cycle link generates 1 result, Fig. 6.5n.
9. For the generalized chain shown in Fig. 6.4i, the assignment of the Olympiad cycle link generates zero results.

Therefore, after the assignment of the Olympiad cycle link subject to the required design constraints, 14 specialized chains with identified ground link, Callippic cycle link, and Olympiad cycle link are available, as shown in Fig. 6.5a–n.

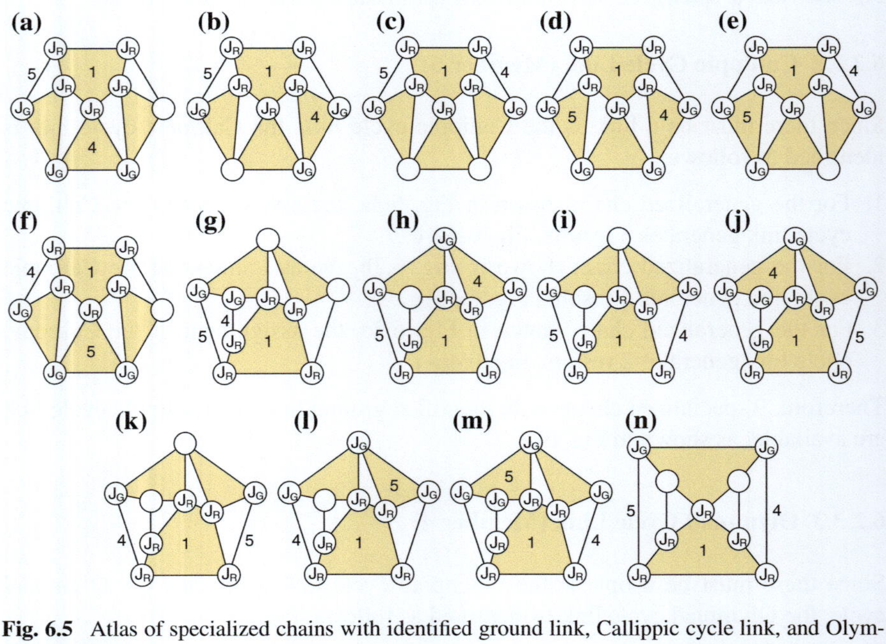

Fig. 6.5 Atlas of specialized chains with identified ground link, Callippic cycle link, and Olympiad cycle link

6.2.3.4 Input Link (Member 2)

Since there must be an input link that at least includes a gear joint and a revolute joint, the input link is identified as follows:

1. For the generalized chain shown in Fig. 6.5a, the assignment of the input link generates 1 result, Fig. 6.6a.
2. For the generalized chain shown in Fig. 6.5b, the assignment of the input link generates zero results.
3. For the generalized chain shown in Fig. 6.5c, the assignment of the input link generates 1 result, Fig. 6.6b.
4. For the generalized chain shown in Fig. 6.5d, the assignment of the input link generates zero results.
5. For the generalized chain shown in Fig. 6.5e, the assignment of the input link generates zero results.
6. For the generalized chain shown in Fig. 6.5f, the assignment of the input link generates 1 result, Fig. 6.6c.
7. For the generalized chain shown in Fig. 6.5g, the assignment of the input link generates 2 results, Fig. 6.6d–e.
8. For the generalized chain shown in Fig. 6.5h, the assignment of the input link generates 1 result, Fig. 6.6f.
9. For the generalized chain shown in Fig. 6.5i, the assignment of the input link generates 1 result, Fig. 6.6g.
10. For the generalized chain shown in Fig. 6.5j, the assignment of the input link generates zero result.
11. For the generalized chain shown in Fig. 6.5k, the assignment of the input link generates 1 result, Fig. 6.6h.
12. For the generalized chain shown in Fig. 6.5l, the assignment of the input link generates 1 result, Fig. 6.6i.

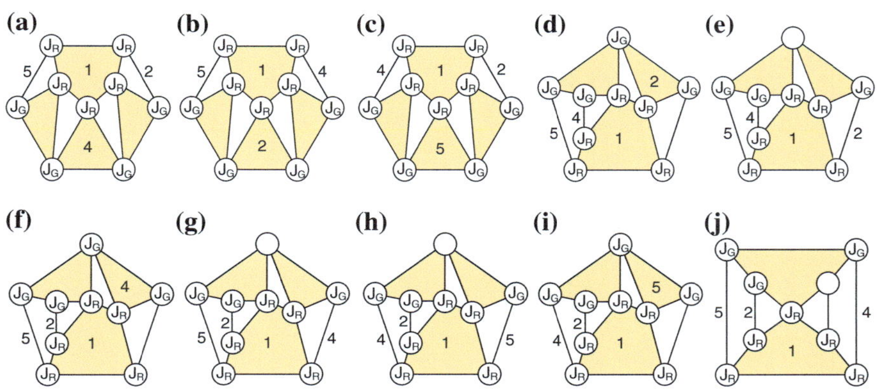

Fig. 6.6 Atlas of specialized chains with identified ground link, Callippic cycle link, Olympiad cycle link, and input link

13. For the generalized chain shown in Fig. 6.5m, the assignment of the input link generates zero results.
14. For the generalized chain shown in Fig. 6.5n, the assignment of the input link generates 1 result, Fig. 6.6j.

Hence, after the assignment of the input link subject to the required design constraints, 10 specialized chains with identified ground link, Callippic cycle link, Olympiad cycle link, and input link are available, Fig. 6.6a–j.

6.2.3.5 Metonic Cycle Link (Member 3)

Since there must be a link as the Metonic cycle link to rotate at the rate of the Metonic cycle, the Metonic cycle link is identified as follows:

1. For the generalized chain shown in Fig. 6.6a, the assignment of the Metonic cycle link generates 1 result, Fig. 6.7a.
2. For the generalized chain shown in Fig. 6.6b, the assignment of the Metonic cycle link generates 1 result, Fig. 6.7b.
3. For the generalized chain shown in Fig. 6.6c, the assignment of the Metonic cycle link generates zero results.
4. For the generalized chain shown in Fig. 6.6d, the assignment of the Metonic cycle link generates 1 result, Fig. 6.7c.
5. For the generalized chain shown in Fig. 6.6e, the assignment of the Metonic cycle link generates 1 result, Fig. 6.7d.
6. For the generalized chain shown in Fig. 6.6f, the assignment of the Metonic cycle link generates 1 result, Fig. 6.7e.

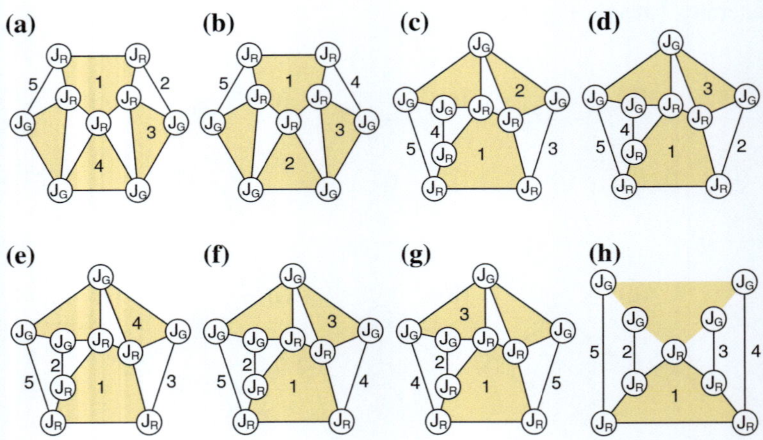

Fig. 6.7 Atlas of specialized chains with identified ground link, Callippic cycle link, Olympiad cycle link, input link, and Metonic cycle link

7. For the generalized chain shown in Fig. 6.6g, the assignment of the Metonic cycle link generates 1 result, Fig. 6.7f.
8. For the generalized chain shown in Fig. 6.6h, the assignment of the Metonic cycle link generates 1 result, Fig. 6.7g.
9. For the generalized chain shown in Fig. 6.6i, the assignment of the Metonic cycle link generates zero results.
10. For the generalized chain shown in Fig. 6.6j, the assignment of the Metonic cycle link generates 1 result, Fig. 6.7h.

Therefore, after the assignment of the Metonic cycle link subject to the required design constraints, 8 specialized chains with identified ground link, Callippic cycle link, Olympiad cycle link, input link, and Metonic cycle link are available, as shown in Fig. 6.7a–h.

6.2.3.6 Transmission Link (Link 6)

Since there must be at least a transmission link incident to a gear joint and a revolute joint, the transmission link is identified as follows:

1. For the generalized chain shown in Fig. 6.7a, the assignment of the Metonic cycle link generates 1 result, Fig. 6.8a.
2. For the generalized chain shown in Fig. 6.7b, the assignment of the Metonic cycle link generates zero results.
3. For the generalized chain shown in Fig. 6.7c, the assignment of the Metonic cycle link generates zero results.
4. For the generalized chain shown in Fig. 6.7d, the assignment of the Metonic cycle link generates 1 result, Fig. 6.8b.
5. For the generalized chain shown in Fig. 6.7e, the assignment of the Metonic cycle link generates zero results.
6. For the generalized chain shown in Fig. 6.7f, the assignment of the Metonic cycle link generates zero results.
7. For the generalized chain shown in Fig. 6.7g, the assignment of the Metonic cycle link generates 1 result, Fig. 6.8c.
8. For the generalized chain shown in Fig. 6.7h, the assignment of the Metonic cycle link generates zero results.

Therefore, after the assignment of the transmission link subject to the required design constraints, 3 specialized chains with identified ground link, Callippic cycle link, Olympiad cycle link, input link, Metonic cycle link, and transmission link are available, as shown in Fig. 6.8. Furthermore, the rotation directions of some specific links of these obtained specialized chains must be examined so as to generate feasible reconstruction designs with the appropriate rotation directions. Here, the Olympiad cycle link has to rotate clockwise, since it is already confirmed that the inscription sequence of the Olympiad dial is shown with clockwise rotation. However, the Olympiad cycle link of the specialized chain shown in Fig. 6.8b

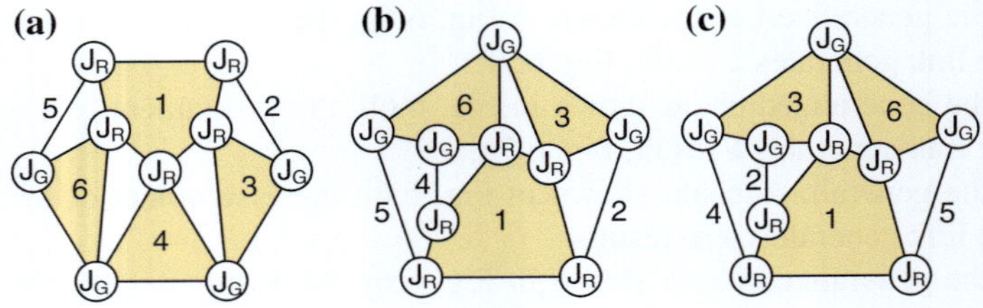

Fig. 6.8 Atlas of specialized chains of the calendrical subsystem

rotates counterclockwise. This specialized chain does not satisfy the constraints of the rotation directions of the Olympiad cycle links. And, the specialized chains shown in Fig. 6.8a, c satisfy the direction constraint of the Olympiad cycle link.

The present studies indicate that the inscriptions on the Callippic dial were not decoded completely and its existence was speculated [3]. The rotation direction of the Callippic cycle link is unconfirmed. Thus, the remaining two specialized chains, as shown in Fig. 6.8a, c, are the feasible specialized chains of the calendrical subsystem under the available decoding information.

6.2.4 *Reconstruction Designs*

Through the process of particularization, the two feasible specialized chains shown in Fig. 6.8a, c are further particularized into their corresponding mechanical schemes, as shown in Fig. 6.9. The design shown in Fig. 6.9a is a new design. Its Metonic cycle link rotates clockwise, and its Olympiad cycle link and Callippic cycle link rotate counterclockwise. The design shown in Fig. 6.9b is the design by Freeth et al. Its Metonic cycle link and Callippic cycle link rotate clockwise, and its Olympiad cycle link rotates counterclockwise. The obvious difference between these two designs is the rotation direction of the Callippic cycle link. Undoubtedly, either one of these two reconstruction designs could be the original mechanism of the calendrical subsystem, because of the unclear inscriptions of the Callippic subsidiary dial. In addition, they are the designs with the simplest mechanisms to generate the required rotation rates.

6.2.4.1 Tooth Calculation of the Feasible Designs

Based on the kinematic analysis of ordinary gear trains, the existing gears, and the known astronomical cycles, the tooth calculation work of the unclear gears of the feasible reconstruction designs can be derived. However, in order to express clearly during the conclusion process of the relations of teeth number for each feasible design, a systematic tooth representation is defined before calculating the teeth of these unclear gears, as follows:

$$T_{\text{gear on the assigned link}}^{\text{gear joint between two adjacent links}}$$

where the superscript denotes the gear joint between two adjacent links and the subscript denotes the gear on the assigned link.

By the known gear ratios, the surviving evidence, and the size constraints, the teeth numbers of these two feasible reconstruction designs are analyzed as follows:

Feasible Reconstruction Design of Fig. 6.9a

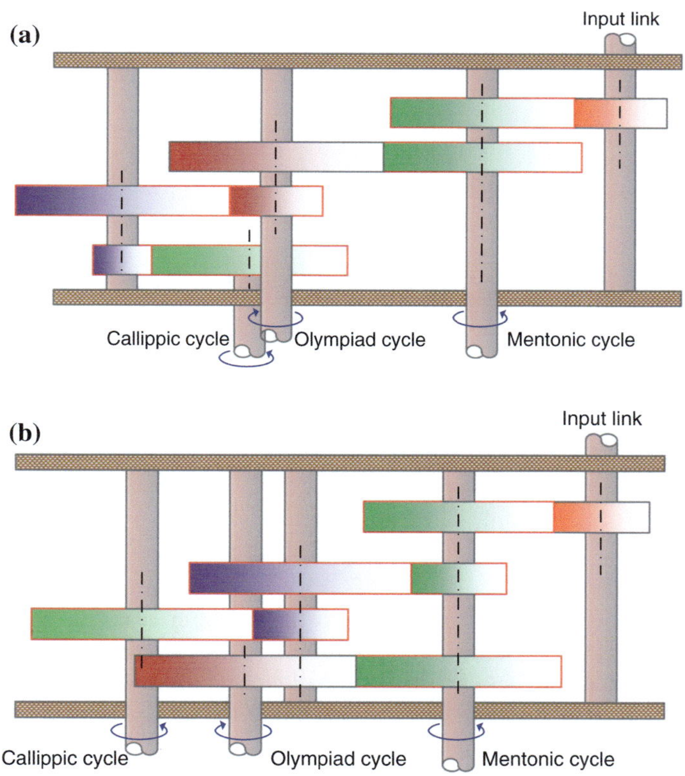

Fig. 6.9 Mechanical drawings of two feasible reconstruction designs of the calendrical subsystem. **a** An alternative design. **b** Available design by Freeth et al.

Based on the relation between the input link and the Metonic cycle link, the following expression is true:

$$\left(\frac{53}{19 \times 3}\right) \times \frac{T_2^{23}}{T_3^{23}} = \frac{5}{19}$$

or

$$\frac{T_3^{23}}{T_3^{23}} = \frac{15}{53} \tag{6.7}$$

Based on the relation between the Metonic cycle link and the Olympiad cycle link, the following expression is true:

$$\frac{5}{19} \times \frac{T_3^{34}}{T_4^{34}} = \frac{1}{4}$$

or

$$\frac{T_3^{34}}{T_4^{34}} = \frac{19}{20} \tag{6.8}$$

Based on the relation between the Olympiad cycle link and the Callippic cycle link, the following expression is true:

$$\frac{1}{4} \times \frac{T_4^{46}}{T_6^{46}} \times \frac{T_6^{56}}{T_5^{56}} = \frac{1}{76}$$

or

$$\frac{T_4^{46}}{T_6^{46}} \times \frac{T_6^{56}}{T_5^{56}} = \frac{1}{19} \tag{6.9}$$

Equations (6.7)–(6.9) show the relations of all meshing gears in the reconstruction designs. Absolutely, all teeth combinations of the reconstructed ordinary gear trains can be calculated based on these equations and the known teeth. And, these concluded equations are necessary for generating the detail design of reconstruction models.

Feasible Reconstruction Design of Fig. 6.9b

The relation between the input link and the Metonic cycle link and the relation between the Metonic cycle link and the Olympiad cycle link are the same as those mentioned above. As to the relation between the Metonic cycle link and the Callippic cycle link, the following expression is true:

$$\frac{5}{19} \times \frac{T_3^{36}}{T_6^{36}} \times \frac{T_6^{56}}{T_5^{56}} = \frac{1}{76}$$

or

$$\frac{T_3^{36}}{T_6^{36}} \times \frac{T_6^{56}}{T_5^{56}} = \frac{1}{20} \qquad (6.10)$$

Therefore, in accordance with Eqs. (6.7), (6.8), and (6.10), all possible teeth combinations of this reconstruction design as shown in Fig. 6.9b are calculated. In addition, the teeth combination of the design by Freeth et al. is also satisfied. The concluded equations are applied to build the reconstruction model of the complete interior mechanism in Chap. 11.

6.3 Remarks

In summary, the design procedure of the calendrical subsystem is illustrated in detail. Two feasible reconstruction designs are generated, including the design by Freeth et al. Except to match the display directions of the Metonic dial and the Olympiad subsidiary dial, these two reconstruction designs, respectively, indicate that the display direction of the Callippic subsidiary dial could be clockwise and counterclockwise.

Furthermore, based on the surviving evidence and the kinematic analysis of gear trains, the tooth calculation relations for all gears of the two feasible reconstruction designs are derived. Once the inscriptions of the Callippic subsidiary dial are decoded in the future, these two designs are helpful to restore the original mechanism of the calendrical subsystem.

References

1. Freeth T et al (2006) Decoding the ancient greek astronomical calculator known as the antikythera mechanism. Nature 444:587–891
2. Freeth T (2009) Decoding an ancient computer. Sci Am 301(6):79–83
3. Freeth T et al (2008) Calendars with olympiad display and eclipse prediction on the antikythera mechanism. Nature 454:614–617
4. Yan HS (2007) Reconstruction designs of lost ancient Chinese machinery. Springer, Netherland
5. Yan HS (1992) A methodology for creative mechanism design. Mech Mach Theory 27(3):235–242
6. Yan HS (1998) Creative design of mechanical devices. Springer, Berlin
7. Yan HS, Lin JL (2011) Reconstruction synthesis of the calendrical subsystem of the Antikythera mechanism. ASME Trans J Mech Des 133(2):021004

Chapter 7
Reconstruction Designs of the Lunar Subsystem

The lunar subsystem is a special and complicated mechanism in the surviving evidence. Various reconstruction designs were presented in the past years. The decoding of this subsystem started from the discovery of a pin-in-slot joint by M.T. Wright who claimed and restored the mechanism of the lunar anomaly motion in the Antikythera device [1–3]. In 2006, Freeth et al. presented a hypothetical gear, which connects the lunar subsystem and the calendrical subsystem. Due to this hypothetical gear, the design concept that there were two input sources to drive the lunar subsystem was presented in their reconstruction design [4–6]. Thus, the output lunar motion satisfies completely the Hipparchus' lunar theory. At present, the reconstruction design by Freeth et al. is the most popular design even though an obvious pentagon pattern detected in the surviving fragment of the lunar subsystem still goes without a proper explanation. This chapter focuses on decoding this unexplained feature of the lunar subsystem.

The ancient lunar theory and the surviving evidence are introduced in Chaps. 1 and 3. Here, kinematic analysis of the Hipparchus' lunar theory is proposed to conclude the design constraints. Through illustrating the design procedure of the lunar subsystem, all feasible reconstruction designs are obtained systematically. In addition, the possible resolutions for the unexplained feature are presented according to these reconstruction designs [7–11].

7.1 Historical Archives of the Lunar Subsystem

The lunar subsystem is confirmed to demonstrate the cyclic motion of the Moon. All gears included in this subsystem were identified clearly from the excavation. Unlike the historical archives of the calendrical subsystem, the mechanism analysis of gear trains to conclude the numbers of members and joints of all

J.-L. Lin and H.-S. Yan, *Decoding the Mechanisms of Antikythera Astronomical Device*, DOI 10.1007/978-3-662-48447-0_7

possible topological structure of mechanisms is unnecessary. This chapter focuses on generating all feasible designs to propose the uncertain pattern of this subsystem based on the known members and joints. In order to define the required design constraints, the kinematic analysis related to the surviving evidence and the applied astronomical theory of the lunar subsystem should be proceeded.

7.1.1 Kinematic Analysis of the Lunar Theory

The design concept applied in the lunar subsystem is on the basis of the development of the astronomy in the era of Antikythera device. As mentioned in Chap. 1, the astronomical theories describing the cyclic motions of celestial bodies were developed highly at that time. Except to inheriting the knowledge of the Babylonian mathematical astronomy, the geometrical astronomy was gradually formed in ancient Greece. The epicyclic system and the eccentric system of the geometrical astronomy had been used to successfully demonstrate the cyclic motion of the Moon. The related astronomical achievements were proposed by Apollonius and Hipparchus [12].

Based on the research of Freeth et al., the lunar subsystem depends on the Hipparchus' lunar theories, including the epicyclic system and the eccentric system. It is proved that these two astronomical systems are equivalent. Therefore, the kinematic analysis of the epicyclic system of the Hipparchus' lunar theory is introduced in detail to conclude the output kinematic characteristics of the system.

From the viewpoint of kinematic analysis, the epicyclic system of the lunar motion described by Hipparchus is regarded as the superimposition of two rotations of the epicycle and the deferent, as shown in Fig. 7.1. To merely consider the rotation of the planet on the epicycle, i.e., when center C is fixed, the following expression to satisfy the geometrical relations is true:

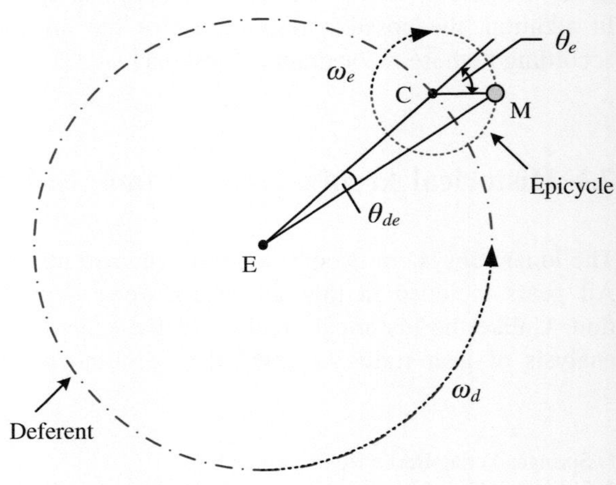

Fig. 7.1 Illustration of the epicyclic system of Hipparchus' lunar theory

$$\frac{r_e}{\sin\theta_{\text{de}}} = \frac{r_d}{\sin(\theta_e - \theta_{\text{de}})}$$

or

$$\sin\theta_{\text{de}} = \frac{r_e}{r_d}\sin\left(\theta_e - \theta_{\text{de}}\right) \tag{7.1}$$

where r_e and r_d are the radii of the epicycle and the deferent, respectively; θ_e and θ_{de} are the angular displacements of the planet for center C and the Earth, respectively.

Since r_e / r_d for each planet is too small, $\sin\theta_{\text{de}}$ approaches θ_{de}. Then, Eq. (7.1) is simplified as follows:

$$\theta_{\text{de}} = \frac{r_e}{r_d}\sin\left(\theta_e - \theta_{\text{de}}\right) \tag{7.2}$$

By differentiating Eq. (7.2) with time, the following relation of the angular rates (ω_{de} and ω_e) about the different references is concluded:

$$\omega_{\text{de}} = \left(\frac{\frac{r_e}{r_d}\cos\left(\theta_e - \theta_{\text{de}}\right)}{1 + \frac{r_e}{r_d}\cos\left(\theta_e - \theta_{\text{de}}\right)}\right)\omega_e = F\omega_e \tag{7.3}$$

where ω_e and ω_{de} are the angular rates of planet about center C of the epicycle and the Earth, respectively, and F is the harmonic anomaly function in the epicyclic system.

In the epicyclic system, center C of the epicyclic should rotate around the Earth with angular rate ω_d. Therefore, while rotation of center C is added, the whole planetary motion relative to the Earth model is generated. In addition, the absolute angular rate of planet about the Earth (ω) is concluded as follows:

$$\omega = \omega_d + \omega_{\text{de}} = \omega_d + \left(\frac{\frac{r_e}{r_d}\cos\left(\theta_e - \theta_{\text{de}}\right)}{1 + \frac{r_e}{r_d}\cos\left(\theta_e - \theta_{\text{de}}\right)}\right)\omega_e = \omega_d + F\omega_e \tag{7.4}$$

Parameters ω_d, ω_e, r_e, and r_d mentioned above are constants, and they are determined by the lunar theory and the records of lunar motion. F is the anomaly function of the epicyclic system. In accordance with Eq. (7.4), this epicyclic system of the Moon, i.e., the second lunar theory, is able to introduce the motion of the Moon with the rate of the sidereal month plus the harmonic variation of the anomalistic month. The variable angular speed curve for the lunar motion can be modeled approximately by the epicyclic gear train with a pin-in-slot device [13, 14]. When two gears incident to the pin-in-slot joint are on the fixed axes, the period of variation is a sidereal motion. Therefore, in order to change the period of variation from the sidereal month to the anomalistic month, it is necessary to mount two gears incident to the pin-in-slot joint on the carrier of the gear train.

7.1.2 Kinematic Analysis of Epicyclic Gear Trains

The surviving evidence definitely indicates that the lunar subsystem is an epicyclic gear train with a pin-in-slot joint, and this joint is incident to the planet gears. In the reconstruction design by Freeth et al., a gear to drive the epicyclic gear is hypothesized as the other input source so as to prove that the lunar subsystem completely demonstrates Hipparchus' lunar theory. The reconstructed lunar subsystem is a 7-bar epicyclic gear train with 10 joints and 2 degrees of freedom. Here, the kinematic analysis of the epicyclic gear train with 7 members and 10 joints is introduced. It is proved that the output characteristic of the epicyclic gear train with a pin-in-slot joint is equal to the output characteristic of the lunar theory.

As mentioned in Chap. 3, the pin-in-slot device is a critical founding of the lunar subsystem to generate the anomaly motion. For the pin-in-slot device shown in Fig. 7.2, the sine law of the trigonometry is as follows:

$$\frac{r}{\sin\theta_s} = \frac{e}{\sin(\theta_p - \theta_s)}$$

or

$$\sin(\theta_p - \theta_s) = \frac{e}{r}\sin\theta_s \tag{7.5}$$

Based on Eq. (7.5), the resulting function of the angle difference between these two rotational members is related to the ratio of the eccentric distance between the rotation centers of the links, in which the pin-in-slot joint is incident, and the pin distance. The surviving evidence indicates that the eccentric distance is about 1.1 (mm), and the pin distance is about 9.6 (mm) [4]. When the eccentric distance (e) is much smaller than the rotational radius of the pin (r), the ratio of these two

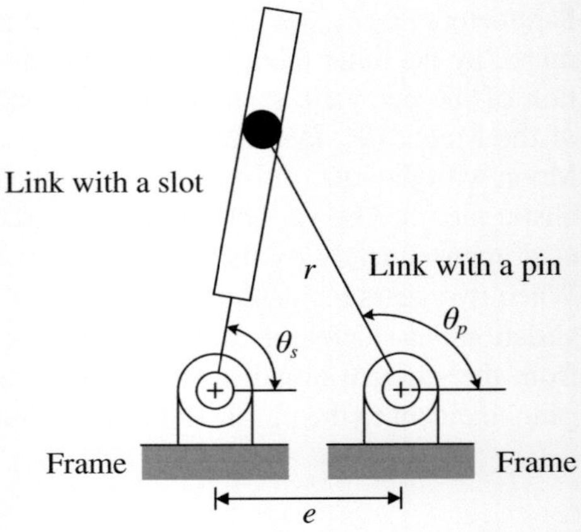

Fig. 7.2 Illustration of a pin-in-slot device

distances is so small that $\sin(\theta_p - \theta_s)$ approximates to $(\theta_p - \theta_s)$. And, Eq. (7.5) is simplified as follows:

$$\theta_p - \theta_s = \frac{e}{r}\sin\theta_s \tag{7.6}$$

Then, by differentiating Eq. (7.6) with time, the relation about the angular speeds of these two links is concluded as follows:

$$\omega_p - \omega_s = \frac{e}{r}\omega_s \cos\theta_s \tag{7.7}$$

or

$$\omega_s = \left[\frac{1}{\left(1 + \frac{e}{r}\cos\theta_s\right)}\right]\omega_p = A\omega_p \tag{7.8}$$

where ω_s is the angular speed of the link with a slot, ω_p is the angular speed of the link with a pin, and A is the harmonic anomaly function resulted by the pin-in-slot joint.

According to Eq. (7.8), it is known that the rotation of the link with a slot (ω_s) should be approximately a variable function at the period of the angular speed of the link with a pin (ω_p), if the angular speed of the link with a pin (ω_p) is constant. Furthermore, observing the illustrations shown in Figs. 7.1 and 7.2, the following expressions are true:

$$\theta_{\mathrm{de}} = \theta_s - \theta_p \tag{7.9}$$

$$\theta_e = \theta_p \tag{7.10}$$

$$r_d = r \tag{7.11}$$

$$r_e = e \tag{7.12}$$

Thus, by submitting Eqs. (7.9)–(7.12) into the anomaly function of the epicyclic system (F), the following equation is obtained:

$$F = \frac{\frac{r_e}{r_d}\cos\left(\theta_e - \theta_{\mathrm{de}}\right)}{1 + \frac{r_e}{r_d}\cos\left(\theta_e - \theta_{\mathrm{de}}\right)} = \frac{\frac{e}{r}\cos\theta_s}{1 + \frac{e}{r}\cos\theta_s} \tag{7.13}$$

To compare the anomaly function of the pin-in-slot device (A) with the anomaly function of the epicyclic system (F), the following expression is concluded:

$$F = 1 - A \tag{7.14}$$

when the eccentric distance (e) is much smaller than the rotational radius of the pin (r), i.e., the radius of the epicycle is much smaller than the radius of the deferent.

In order to demonstrate the lunar motion, the rate analysis of the epicyclic gear train with 10 joints including a pin-in-slot joint should be discussed further. It is confirmed by the surviving evidence that the pin-in-slot joint is incident to the planet gears. Here, to consider an epicyclic gear with n planet gears ($n = 2, 3, \ldots$), it is supposed that the remaining planet gears are adjacent to each other by a gear joint except that the kth and the ($k-1$)th planet gears are adjacent to each other by a pin-in-slot joint ($k = 2, 3, \ldots, n$). The related rate analysis is concluded as follows:

1. Assume all members of the epicyclic gear train rotate at x rpm. Then, the rates of gears are obtained as follows:

$$\omega'_s = \omega'_c = \omega'_{p(k-1)} = \omega'_{\mathrm{pk}} = \omega'_{\mathrm{pn}} = \omega'_o = x \tag{7.15}$$

   where ω'_s, ω'_c, $\omega'_{p(k-1)}$, ω'_{pk}, ω'_{pn}, and ω'_o are, respectively, the rates of the sun gear, the carrier, the kth planet gear, the ($k-1$)th planet gear, the nth planet gear, and the output gear without considering the engaging of gears.
2. The carrier does not contribute to the motions of the epicyclic gear train, i.e., assuming that the sun gear rotates at y rpm, when the carrier is fixed. By the engaging of gears, the rates of gears are obtained as follows:

$$\omega''_s = y \tag{7.16}$$

$$\omega''_c = 0 \tag{7.17}$$

$$\omega''_{p(k-1)} = (-1)^{k-1}\left(\frac{T_s}{T_{p(k-1)}}\right)y \tag{7.18}$$

$$\omega''_{\mathrm{pk}} = (-1)^{k-1}(1-F)\left(\frac{T_s}{T_{p(k-1)}}\right)y \tag{7.19}$$

$$\omega''_{\mathrm{pn}} = (-1)^{n-1}(1-F)\left(\frac{T_s T_{\mathrm{pk}}}{T_{p(k-1)} T_o}\right)y \tag{7.20}$$

$$\omega''_o = (-1)^{n}(1-F)\left(\frac{T_s T_{\mathrm{pk}}}{T_{p(k-1)} T_o}\right)y \tag{7.21}$$

   where ω''_s, ω''_c, $\omega''_{p(k-1)}$, ω''_{pk}, ω''_{pn}, and ω''_o are the rates of the sun gear, the carrier, the kth planet gear, the ($k-1$)th planet gear, the nth planet gear, and the output gear with only the engaging of gears considered, respectively, and T_s, $T_{p(k-1)}$, T_{pk}, and T_o are the teeth of the sun gear, the ($k-1$)th planet gear, the kth planet gear, and the output gear, respectively.
3. Superimposing the two motions mentioned above, the absolute rates are concluded as follows:

$$\omega_s = x + y \tag{7.22}$$

$$\omega_c = x \tag{7.23}$$

$$\omega_{p(k-1)} = x + (-1)^{k-1}\left(\frac{T_s}{T_{p(k-1)}}\right)y \tag{7.24}$$

$$\omega_{\text{pk}} = x + (-1)^{k-1}(1-F)\left(\frac{T_s}{T_{p(k-1)}}\right)y \tag{7.25}$$

$$\omega_{\text{pn}} = x + (-1)^{n-1}(1-F)\left(\frac{T_s T_{\text{pk}}}{T_{p(k-1)}T_o}\right)y \tag{7.26}$$

$$\omega_o = x + (-1)^{n}(1-F)\left(\frac{T_s T_{\text{pk}}}{T_{p(k-1)}T_o}\right)y \tag{7.27}$$

where ω_s, ω_c, $\omega_{p(k-1)}$, ω_{pk}, ω_{pn}, and ω_o are the absolute rates of the sun gear, the carrier, the kth planet gear, the $(k-1)$th planet gear, the nth planet gear, and the output gear, respectively, and A is the quasi-sinusoidal function caused by the pin-in-slot joint.

Based on Eq. (7.14), the following expression about angular speed of the output gear is true:

$$\omega_o = x + (-1)^{n}(1-F)\left(\frac{T_s T_{\text{pk}}}{T_{p(k-1)}T_o}\right)y \tag{7.28}$$

Once the number of planet gears is even, the angular speed of the output gear is obtained as follows:

$$\omega_o = x + (1-F)\left(\frac{T_s T_{\text{pk}}}{T_{p(k-1)}T_o}\right)y$$

or

$$\omega_o = \left(x + \frac{T_s T_{\text{pk}}}{T_{p(k-1)}T_o}y\right) + \left(-\frac{T_s T_{\text{pk}}}{T_{p(k-1)}T_o}\right)y \tag{7.29}$$

According to Eqs. (7.4) and (7.29), the following expressions about the teeth and the rates of deferent and epicycle are true:

$$\omega_d = x + \frac{T_s T_{\text{pk}}}{T_{p(k-1)}T_o}y \tag{7.30}$$

$$\omega_e = \left(-\frac{T_s T_{\text{pk}}}{T_{p(k-1)} T_o} \right) y \tag{7.31}$$

In conclusion, once the rates of the sun gear and the carrier of the epicyclic gear train are known, the teeth could be calculated by Eqs. (7.30) and (7.31). The sun gear and the carrier are the two required input sources of the lunar subsystem. The above viewpoint is confirmed completely in the research by Freeth et al.

7.2 Design Process of the Lunar Subsystem

The research by Freeth et al. indicated one unexplained feature, i.e., a visible mark with the shape of regular pentagon located at the center of gear of the lunar subsystem [4]. This obvious feature would affect the identifications of members of the mechanism. In addition, the arrangement of axes in the design would relate to the manufacturing possibility in the ancient time. In order to obtain alternative concepts to clarify the unexplained feature from the feasible reconstruction designs, the design procedure of the lunar subsystem is demonstrated.

7.2.1 Design Specifications

The historical archives of the lunar subsystem include the surviving evidence, the topological analysis of the existing designs, and the kinematic analysis of the lunar theory. In accordance with these studies, the design specifications are concluded as follows:

1. It is a planar 7-bar mechanism with 10 joints.
2. The design requires two input sources to generate the specific output of the members, i.e., 2 degrees of freedom.
3. The types of mechanical members are external gears and links.
4. The joints consist of 6 revolute joints, a cam joint (pin-in-slot joint), and 3 external gear joints.

7.2.2 Generalized Chains

According to the concept of generalization, the generalized chain with 7 links and 10 joints corresponding to the design by Freeth et al. as shown in Fig. 4.9 is obtained, as shown in Fig. 7.3. Then, based on the concept of number synthesis, there are 50 generalized chains with 7 links and 10 joints [9]. However, not all these generalized chains are feasible for the lunar subsystem. Based on the

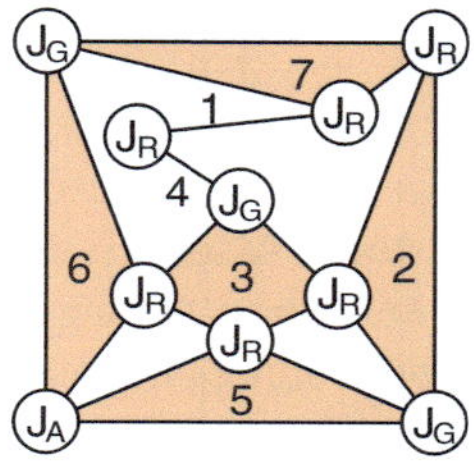

Fig. 7.3 Generalized chain corresponding to the design by Freeth et al.

analysis of the epicyclic gear train with a pin-in-slot device as mentioned in Sect. 4.5, it is known that feasible generalized chains must have a 3-bar loop formed exclusively by multiple links. The link with a pin and the link with a slot must be installed with a common link. Furthermore, in order to satisfy the constraints of joint distributions, two ternary links of the 3-bar loop must not be composed of other 3-bar loops.

The existing design presented a hypothetical input to generate the carrier with the rate of anomalistic month. For the lunar subsystem, the assumption is the key to perfectly demonstrate Hipparchus' lunar theory. Therefore, there must be one binary link adjacent to the multiple link of the 3-bar loop in accordance with the assumption of the input. Here, based on the above constraints, only 13 generalized chains with 7 links and 10 joints are feasible, as shown in Fig. 7.4.

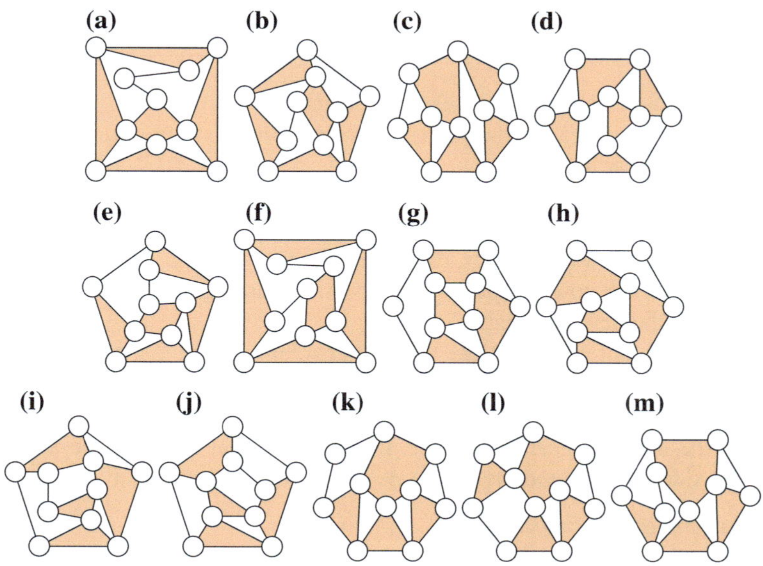

Fig. 7.4 Atlas of feasible generalized chains with 7 links and 10 joints

7.2.3 Specialized Chains

Kinematic analyses of the Hipparchus' lunar theory and the epicyclic gear trains with a pin-in-slot joint are understood. Also, it is proved that the anomaly function of the epicyclic system is much related to the anomaly function of the pin-in-slot joint. Here, on the basis of the above kinematic analyses and the previous study of the related historical records, the ancient astronomy, and the existing designs, the design constraints of the lunar subsystem are concluded as follows:

1. The input with the rotation rate of the sidereal month and the output must be coaxial in order to satisfy the space arrangement according to the object evidence.
2. The lunar subsystem must have the function of variable transmission to demonstrate the lunar motion in accordance with Hipparchus's first lunar theory.
3. The epicyclic gear train must at least include 1 ground link, 1 sidereal gear, 1 deferent gear, 1 pin-in-slot device, 1 anomalistic gear, and 1 output gear. Design constraints of each member are stated as follows:

   Ground link (member 1, K_F)

   (a) There must be 1 ground link as the frame.
   (b) The ground link must not be adjacent to the planet gears.

   Sidereal link (member 2, K_I)

   (a) There must be 1 sidereal gear as the input link.
   (b) The sidereal gear is the input with the rotation rate of the sidereal month.
   (c) The sidereal gear is adjacent to the pin-in-slot device with a gear joint.

   Deferent link (member 3, K_D)

   (a) There must be 1 deferent link as the carrier.
   (b) The deferent gear is the carrier corresponding to each planet gear of the reverted gear train.
   (c) The deferent gear must be a quaternary link at least.
   (d) When the deferent gear, the sidereal gear, and the output gear are adjacent to each other successively, the deferent gear is one of the coaxial links. In addition, the power input/output and the planet gears are in the opposite side of the deferent gear.
   (e) When the deferent gear, the sidereal gear, and the output gear are not adjacent to each other successively, the deferent gear is not one of the coaxial links and the power input/output and the planet gears are in the same side of the deferent gear. Therefore, the deferent gear must not be adjacent to the sidereal gear and the output gear to avoid the redundant axis structure.

   Anomalistic link (member 4, K_A)
   As far as the lunar subsystem is concerned, another rotation rate is necessary to achieve the simulation of lunar anomaly motion, not including the confirmed rate

of the sidereal month. The purpose of the anomalistic gear is to transmit the power to the deferent gear and to generate the variation period of the anomalistic month.

(a) The anomalistic gear is a binary link.
(b) The anomalistic gear is adjacent to the ground link with a revolute joint and to the deferent gear with a gear joint.
(c) The anomalistic gear must rotate counterclockwise.

Pin-in-slot device (member 5, K_P; member 6, K_S)
The pin-in-slot device consisting of the pin gear (member 5) and the slot gear (member 6) is able to generate a variable speed transmission to satisfy the lunar theory. Furthermore, previous analysis points out the following related constraints:

(a) The pin gear and the slot gear are ternary links.
(b) The pin gear and the slot gear are adjacent to the deferent gear with a revolute joint, respectively.
(c) The pin gear is adjacent to the slot gear with a cam joint in order to generate the function of variable rotation rate.

Output link (member 7, K_O)

(a) There must be 1 link as the output gear.
(b) The output gear is adjacent to the pin-in-slot device with a gear joint.

Revolute joint (joints a, b, c, e, g, and h; J_R)

(a) Every link must have at least 1 revolute joint.
(b) There can be no loop formed exclusively by revolute joints in order to avoid the generation of rigid chains, i.e., subchains with zero degrees of freedom.
(c) Any joint incident to the ground link must be a revolute joint.

Gear joint (joints d, f, and j; J_G)

(a) There must be 3 gear joints.
(b) No 3-bar loop can be formed exclusively by gear joints, since the epicyclic gear train is formed by simple gears.
(c) Except for the ground link, the remaining links must have at least 1 gear joint.

Since the research target is to decode the unexplained feature on the deferent link, the relations between the coaxial links, including the deferent link, are discussed. Figure 5.13 clearly indicates the arrangement of axes for the coaxial links, i.e., members 2, 3, and 7. This design regarded a multiple joint is incident to these coaxial links. Such a viewpoint avoids generating any equivalent coaxial design. Furthermore, through the positions of the input power, the output power, and the planet gears corresponding to the deferent link, the possible relations between the coaxial links and the deferent link are analyzed. If the sources of the input power and the output power, as well as the planet gears are on the opposite side of the deferent gear, the power transmission should go through the deferent link.

In addition, the multiple joint must be incident to the deferent link. If the sources of the input power and the output power and the planet gears are on the same side of the deferent link, the multiple joint cannot always be incident to the deferent link.

The specialization process of the lunar subsystem is carried out after concluding the appropriate design constraints. Once the feasible kinematic chains corresponding to the lunar subsystem are obtained, all possible specialized chains are identified by the following steps:

Step 1. For each feasible generalized chain, the pin-in-slot device is assigned.
Step 2. For each result from Step 1, the anomalistic gear is assigned.
Step 3. For each result from Step 2, the ground link is assigned.
Step 4. For each result from Step 3, the remaining links, including the sidereal gear and the output gear, are assigned.
Step 5. For each result from Step 4, the revolute joints are assigned.
Step 6. For each result from Step 5, the gear joints are assigned.

Therefore, for the atlas of generalized chains with 7 links and 10 joints shown in Fig. 7.4, all corresponding specialized chains are identified as follows.

7.2.3.1 Pin-in-Slot Device (Members 3, 5, and 6, and Joint J_A)

Since there must be a 3-bar loop including 2 ternary links as the pin-in-slot device and the joint incident to these ternary links must be a pin-in-slot joint, the pin-in-slot device is identified as follows:

1. For the generalized chain shown in Fig. 7.4a, the assignment of the pin-in-slot device generates 1 result, as in Fig. 7.5a.
2. For the generalized chain shown in Fig. 7.4b, the assignment of the pin-in-slot device generates 1 result, as in Fig. 7.5b.
3. For the generalized chain shown in Fig. 7.4c, the assignment of the pin-in-slot device generates 1 result, as in Fig. 7.5c.
4. For the generalized chain shown in Fig. 7.4d, the assignment of the pin-in-slot device generates 1 result, as in Fig. 7.5d.
5. For the generalized chain shown in Fig. 7.4e, the assignment of the pin-in-slot device generates 2 results, as in Fig. 7.5e, f.
6. For the generalized chain shown in Fig. 7.4f, the assignment of the pin-in-slot device generates 1 result, as in Fig. 7.5g.
7. For the generalized chain shown in Fig. 7.4g, the assignment of the pin-in-slot device generates 1 result, as in Fig. 7.5h.
8. For the generalized chain shown in Fig. 7.4h, the assignment of the pin-in-slot device generates 1 result, as in Fig. 7.5i.
9. For the generalized chain shown in Fig. 7.4i, the assignment of the pin-in-slot device generates 1 result, as in Fig. 7.5j.
10. For the generalized chain shown in Fig. 7.4j, the assignment of the pin-in-slot device generates 1 result, as in Fig. 7.5k.

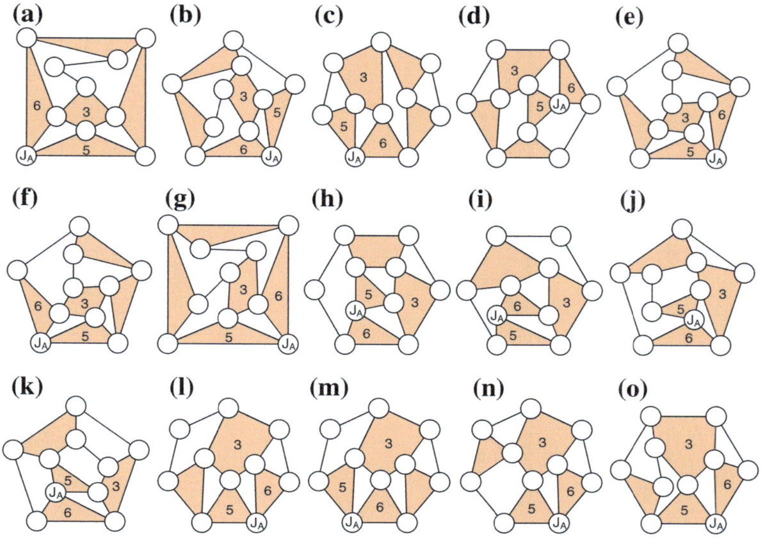

Fig. 7.5 Atlas of specialized chains with identified pin-in-slot device

11. For the generalized chain shown in Fig. 7.4k, the assignment of the pin-in-slot device generates 2 results, as in Fig. 7.5l, m.
12. For the generalized chain shown in Fig. 7.4l, the assignment of the pin-in-slot device generates 1 result, as in Fig. 7.5n.
13. For the generalized chain shown in Fig. 7.4m, the assignment of the pin-in-slot device generates 1 result, as in Fig. 7.5o.

Therefore, 15 specialized chains with identified pin-in-slot device are available, as shown in Fig. 7.5a–o.

7.2.3.2 Anomalistic Link (Member 4)

Since there must be an anomalistic link that is a binary link and adjacent to the ground link, the anomalistic link is identified as follows:

1. For the generalized chain shown in Fig. 7.5a, the assignment of the anomalistic link generates 1 result, as in Fig. 7.6a.
2. For the generalized chain shown in Fig. 7.5b, the assignment of the anomalistic link generates 1 result, as in Fig. 7.6b.
3. For the generalized chain shown in Fig. 7.5c, the assignment of the anomalistic link generates zero results.
4. For the generalized chain shown in Fig. 7.5d, the assignment of the anomalistic link generates 1 result, as in Fig. 7.6c.

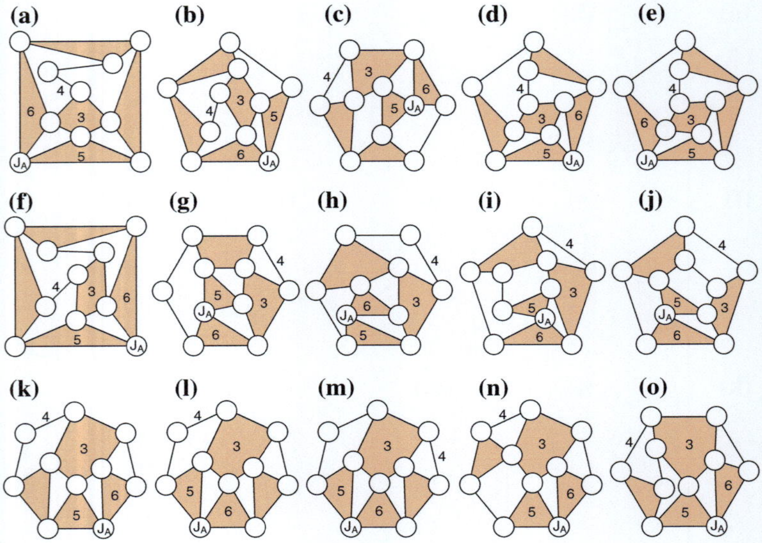

Fig. 7.6 Atlas of specialized chains with identified pin-in-slot device and anomalistic link

5. For the generalized chain shown in Fig. 7.5e, the assignment of the anomalistic link generates 1 result, as in Fig. 7.6d.
6. For the generalized chain shown in Fig. 7.5f, the assignment of the anomalistic link generates 1 result, as in Fig. 7.6e.
7. For the generalized chain shown in Fig. 7.5g, the assignment of the anomalistic d link generates 1 result, as in Fig. 7.6f.
8. For the generalized chain shown in Fig. 7.5h, the assignment of the anomalistic link generates 1 result, as in Fig. 7.6g.
9. For the generalized chain shown in Fig. 7.5i, the assignment of the anomalistic link generates 1 result, as in Fig. 7.6h.
10. For the generalized chain shown in Fig. 7.5j, the assignment of the anomalistic link generates 1 result, as in Fig. 7.6i.
11. For the generalized chain shown in Fig. 7.5k, the assignment of the anomalistic link generates 1 result, as in Fig. 7.6j.
12. For the generalized chain shown in Fig. 7.5l, the assignment of the anomalistic link generates 1 result, as in Fig. 7.6k.
13. For the generalized chain shown in Fig. 7.5m, the assignment of the anomalistic link generates 2 results, as in Fig. 7.6l, m.
14. For the generalized chain shown in Fig. 7.5n, the assignment of the anomalistic link generates 1 result, as in Fig. 7.6n.
15. For the generalized chain shown in Fig. 7.5o, the assignment of the anomalistic link generates 1 result, as in Fig. 7.6o.

Therefore, 15 specialized chains with identified pin-in-slot device and anomalistic link are available, as shown in Fig. 7.6a–o.

7.2.3.3 Ground Link (Member 1)

A ground link must be a multiple link. And, the 2 ternary links, which are included in the 3-bar loop and incident to the pin-in-slot joint, must not be adjacent to the ground link. Therefore, the ground link is identified as follows:

1. For the generalized chain shown in Fig. 7.6a, the assignment of the ground link generates 1 result, as in Fig. 7.7a.
2. For the generalized chain shown in Fig. 7.6b, the assignment of the ground link generates zero results.
3. For the generalized chain shown in Fig. 7.6c, the assignment of the ground link generates 1 result, as in Fig. 7.7b.
4. For the generalized chain shown in Fig. 7.6d, the assignment of the ground link generates zero results.
5. For the generalized chain shown in Fig. 7.6e, the assignment of the ground link generates 1 result, as in Fig. 7.7c.
6. For the generalized chain shown in Fig. 7.6f, the assignment of the ground link generates zero results.
7. For the generalized chain shown in Fig. 7.6g, the assignment of the ground link generates zero results.
8. For the generalized chain shown in Fig. 7.6h, the assignment of the ground link generates 1 result, as in Fig. 7.7d.
9. For the generalized chain shown in Fig. 7.6i, the assignment of the ground link generates 1 result, as in Fig. 7.7e.
10. For the generalized chain shown in Fig. 7.6j, the assignment of the ground link generates zero results.
11. For the generalized chain shown in Fig. 7.6k, the assignment of the ground link generates 1 result, as in Fig. 7.7f.

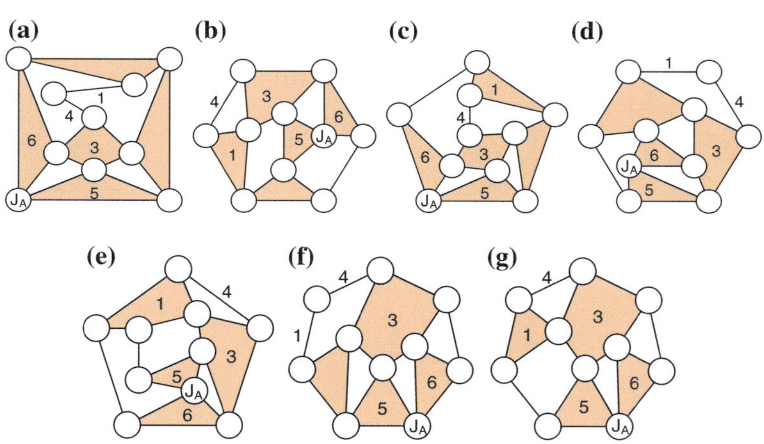

Fig. 7.7 Atlas of specialized chains with identified pin-in-slot device, anomalistic link, and ground link

12. For the generalized chain shown in Fig. 7.6l, the assignment of the ground link generates zero results.
13. For the generalized chain shown in Fig. 7.6m, the assignment of the ground link generates zero results.
14. For the generalized chain shown in Fig. 7.6n, the assignment of the ground link generates 1 result, as in Fig. 7.7g.
15. For the generalized chain shown in Fig. 7.6o, the assignment of the ground link generates zero results.

Therefore, 7 specialized chains with identified pin-in-slot device, anomalistic link, and ground link are available, as shown in Fig. 7.7a–g.

7.2.3.4 Sidereal Link and Output Link (Members 2 and 7)

Since there must be a binary link and a ternary link, respectively, as sidereal link and output link, the sidereal link and the output link are identified as follows:

1. For the generalized chain shown in Fig. 7.7a, the assignments of the sidereal link and the output link generate 2 results, as in Fig. 7.8a, b.
2. For the generalized chain shown in Fig. 7.7b, the assignments of the sidereal link and the output link generate 2 results, as in Fig. 7.8c, d.
3. For the generalized chain shown in Fig. 7.7c, the assignments of the sidereal link and the output link generate zero results.
4. For the generalized chain shown in Fig. 7.7d, the assignments of the sidereal link and the output link generate 2 results, as in Fig. 7.8e, f.
5. For the generalized chain shown in Fig. 7.7e, the assignments of the sidereal link and the output link generate 1 result, as in Fig. 7.8g.
6. For the generalized chain shown in Fig. 7.7f, the assignments of the sidereal link and the output link generate 2 results, as in Fig. 7.8h, i.

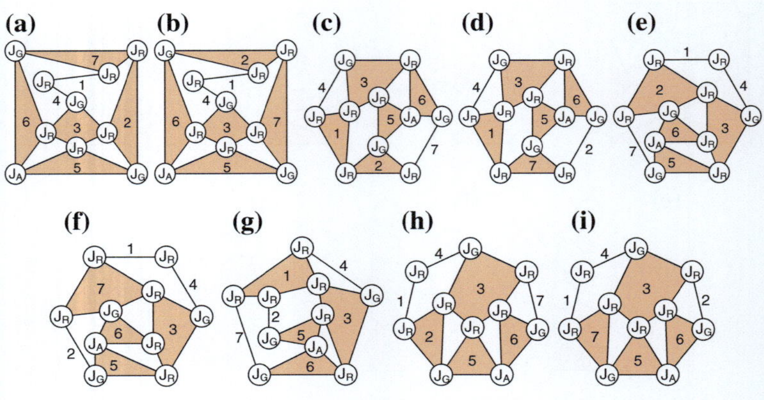

Fig. 7.8 Atlas of specialized chains with identified pin-in-slot device, anomalistic link, ground link, revolute joints, and gear joints

7. For the generalized chain shown in Fig. 7.7g, the assignments of the sidereal link and the output link generate zero results.

Therefore, 9 specialized chains with identified pin-in-slot device, anomalistic link, ground link, sidereal link, and output link are available, as in Fig. 7.8a–i.

7.2.3.5 Revolute Joints (Joints J_R)

There must be 6 revolute joints in each design. Each link must be incident to a revolute joint at least. Therefore, all revolute joints are identified for the 9 specialized chains with identified pin-in-slot device, anomalistic link, ground link, sidereal link, and output link, as shown in Fig. 7.8a–i.

7.2.3.6 Gear Joints (J_G)

Since there must be three gear joints in each design, the remaining three unassigned joints in the 9 specialized chains with identified pin-in-slot device, anomalistic link, ground link, sidereal link, output link, and revolute joints are gear joints, as shown in Fig. 7.8a, i.

In conclusion, through each step of assignments subject to the design constraints, there are 9 specialized chains with identified ground link, sidereal gear, anomalistic gear, output gear, pin-in-slot device, revolute joints, and gear joints, as shown in Fig. 7.8a–i.

7.2.4 Reconstruction Designs

The obtained specialized chains shown in Fig. 7.8a, i are particularized into the schematic diagrams and following the geometry constraints derived from the CT (computed tomography) images of the excavation. However, the shaft structure with coaxial links in these designs should be discussed first. The sequence changes of the coaxial links from inner to outer generate different shaft structures. These different shaft structures are regarded as the designs with a multiple joint, which is incident to the deferent link (member 3), the sidereal link (member 2), and the output link (member 7), i.e., these coaxial designs are isomorphic. Therefore, based on the isomorphic concept, the 9 specialized chains shown in Fig. 7.8 are reduced. The 6 specialized chains shown in Fig. 7.8a, b, e, f, h, i are regarded as one equivalent design. In addition, the coaxial links including the sidereal gear (member 2), the deferent gear (member 3), and the output gear (member 7) are connected with a multiple joint. The remaining 3 specialized chains shown in Fig. 7.8c, d, g are regarded as another equivalent design. Their sidereal gears (member 2), the output gears (member 7), and the ground links (member 1) are coaxial links and connected with a multiple joint.

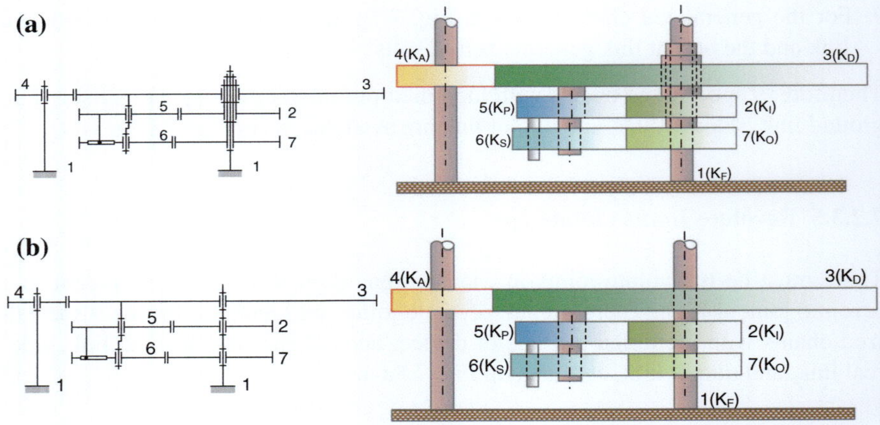

Fig. 7.9 Atlas of the mechanical schematic drawings corresponding to the specialized chains shown in Fig. 7.8. **a** Existing design by Freeth et al. **b** An alternative design

Consequently, by the concepts of coaxial links and multiple joint, only 2 reconstruction designs are feasible, as shown in Fig. 7.9a, b. Here, Fig. 7.9a is the existing design by Freeth et al., and Fig. 7.9b is the other possible design. Figure 7.10 shows these 2 reconstruction designs in 3D solid models.

By observing the reconstruction designs shown in Fig. 7.10, it is understood that different members of axis *E* strongly affect the link identifications for the coaxial gears such as *e*5, *e*6, *e*1, and *e*2 indicated in the design, as shown in Figure 4.9, by Freeth et al. The most important point is that the design of the shaft structure is much related to the unexplained feature with the pentagon pattern. Furthermore, these designs provide alternative viewpoints to discuss the meaning of the unexplained feature. Here, the purposes of the visible pentagonal mark (the explained feature) are proposed in two ways: One is the hub of rotational shaft, i.e., the structure with the inner and the outer shafts existed; another is just a decorative design, i.e., the shaft with the complex structure is unnecessary. In the first explanation, the purpose of the hub design is feasible for both of the obtained reconstruction designs, as shown in Fig. 7.10a, b. Each design can be combined with the remaining mechanisms of Antikythera device to generate appropriate link identifications for the coaxial gears. Some geometrical positions of gears in the space will be changed. Furthermore, in the second explanation (a decorative design), since the design with the inner and the outer shafts is nonexistent, the coaxial gears are not adjacent to each other. Therefore, only the result in Fig. 7.10b can achieve this purpose.

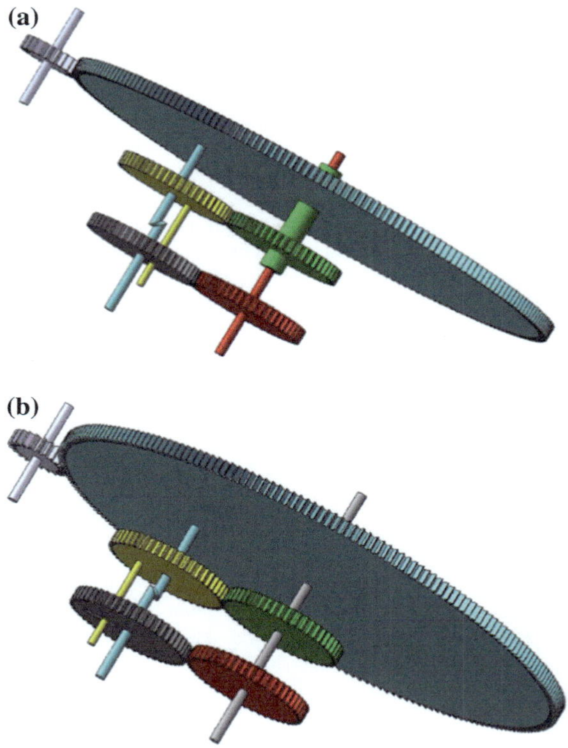

Fig. 7.10 3D models of feasible reconstruction designs of the lunar subsystem. **a** Corresponding to Fig. 7.9a. **b** Corresponding to Fig. 7.9b

7.3 Remarks

In conclusion, the two feasible designs corresponding to the different structures of axis E and the link identifications of the coaxial gears are presented. These designs help to clarify the unexplained feature in the lunar subsystem. Although they have the same output function and satisfy the technology standard of the subject's time period, the complexity of the shaft structure affects the manufacture possibility. Finally, since the teeth of the epicyclic gear train of the lunar subsystem are available from decoding the surviving fragments, the relations of tooth counting for building the reconstruction model are unnecessary.

References

1. Wright MT (2003) Epicyclic gearing and the Antikythera mechanism: part 1. Antiquarian Horology 27:270–279
2. Wright MT (2005) The Antikythera mechanism: a new gearing scheme. Bull Sci Instrum Soc 85:2–7

3. Wright MT (2005) Epicyclic gearing and the Antikythera mechanism: part 2. Antiquarian Horology 29:51–63
4. Freeth T et al (2006) Decoding the ancient greek astronomical calculator known as the Antikythera mechanism. Nature 444:587–891
5. Freeth T (2009) Decoding an ancient computer. Sci Am 301(6):79–83
6. Freeth T et al (2008) Calendars with olympiad display and eclipse prediction on the Antikythera mechanism. Nature 454:614–617
7. Lin JL, Yan HS (2014) Historical development of reconstruction designs of Antikythera mechanism. Appl Mech Mater 163:1–6. (2012 April Presented in Proceedings of the 9th International Conference on History of Mechanical Technology and Mechanical Design, Tainan, 23–25 March 2014)
8. Yan HS (2007) Reconstruction designs of lost ancient Chinese machinery. Springer, Netherland
9. Yan HS (1992) A methodology for creative mechanism design. Mech Mach Theory 27(3):235–242
10. Yan HS (1998) Creative design of mechanical devices. Springer, Singapore
11. Yan HS, Lin JL (2012) Reconstruction synthesis of the unexplained feature in the lunar subsystem of Antikythera mechanism. Proc Inst Mech Eng Part C J Mech Eng Sci 226(C3):1053–1067
12. Linton CM (2004) From Eudoxus to Einstein: a history of mathematical astronomy. Cambridge University, New York
13. Freeth T et al (2006) Supplementary notes. In: Decoding the ancient greek astronomical calculator known as the Antikythera mechanism. Nature 444:587–891
14. Kotsier T (2008) Phase in the unraveling of the secrets of the gear system of the antikythera mechanism. In: Yan HS, Ceccarelli M (eds) Proceedings of HMM 2008. Springer, Netherlands, pp 269–294. International Symposium on History of Machines and Mechanisms, Tainan, 2008

Chapter 8
Reconstruction Designs of the Solar Subsystem

This chapter targets the reconstruction work of the interior mechanism of the lost solar subsystem. The input sources of the lost structures are analyzed according to the surviving structure. The identifications of input sources are the basis of the reconstruction designs. All feasible reconstruction designs are synthesized through the proposed systematical design procedure presented in Chap. 5 [1–5]. In addition, the tooth counting relations corresponding to the feasible designs are derived.

8.1 Historical Archives of the Solar Subsystem

The existence of the solar subsystem is a question because of the losing of corresponding evidence. J. Evans, C.C. Carman, and A.S. Thorndike presented a new design of the front zodiac dial to simulate the solar anomaly motion by the design of an ordinary mechanism [6–8]. However, it is a special viewpoint to decode the display function of the solar motion. In most research, the concept that the partial mechanism of the lost structures (the solar subsystem) generates the solar motion on the zodiac dial with uniform scale is claimed [9–11]. M.T. Wright presented a reconstruction model including the solar and planetary motions. This model is based on his professional knowledge and experience from the reconstruction research of other ancient mechanical devices. Therefore, by the viewpoint of mechanism design and the study of the existing designs, all feasible reconstruction designs of the solar subsystem can be synthesized systematically.

Since the interior structure designated for the solar motion was lost and the relevant historical records were nonexistent, the topological structure of the real mechanism and the input source to operate the solar subsystem are unknown. The types and the numbers of members and joints of the mechanism are uncertain. Hence, the building of the corresponding historical archives is to resolve the above-mentioned problems.

J.-L. Lin and H.-S. Yan, *Decoding the Mechanisms of Antikythera Astronomical Device*, DOI 10.1007/978-3-662-48447-0_8

At first, the possible input arrangements to operate the solar subsystem should be proposed. Then, focusing on the lost mechanism, the only possible way to derive the possible topological structure of mechanism must be acquired through the integrated study of ancient astronomical theories and mechanisms. Here, according to the previous studies on the historical development of Western astronomical theories, Apollonius' two equivalent models and Hipparchus' solar theory are analyzed. Finally, based on the possible input arrangements and the study of astronomy, the characteristics of topological structure of mechanisms are concluded so as to define the design specifications to start the systematic design procedure.

8.1.1 Possible Arrangements of the Driving Power

In order to conclude the characteristics of the topological structures of mechanism of the solar subsystem, the identification work of its driving power sources is necessary. This work not only affects the topological structures of mechanisms but also determines the numbers of teeth corresponding to the resulting designs. By directly observing the mechanism detected from the surviving evidence and studying the existing designs, three possible single driving power arrangements to operate the solar subsystem are suggested, as shown in Fig. 8.1 [5]. Case 1 is that the input link may be connected to the outer shaft along axis *b*. Without the change of rate, it rotates clockwise with the rate of input at the period of a tropical year. Case 2 is that the input link engages with the existing contrate gear *a*1.

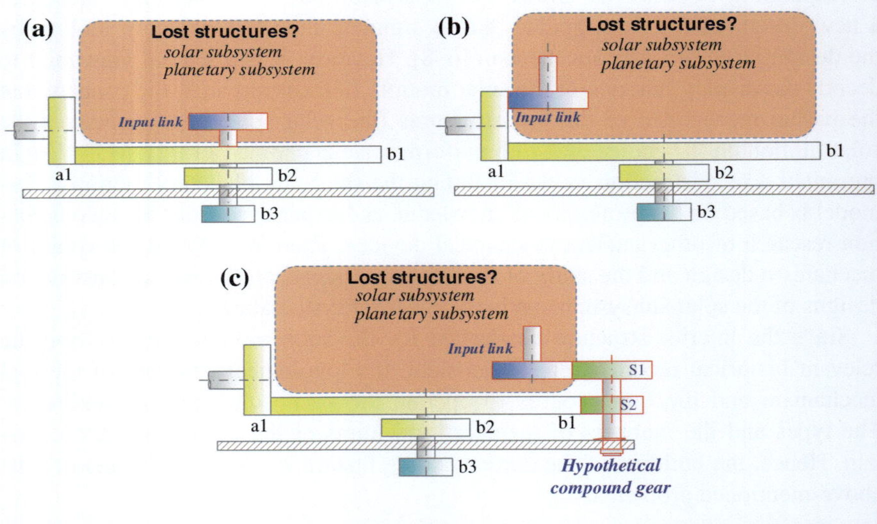

Fig. 8.1 Three possible single driving power arrangements for the lost mechanisms [5]. **a** Case 1, **b** Case 2, **c** Case 3

It rotates counterclockwise. Case 3 is that the input link can rotate clockwise and with different input rates. One compound gear is supposed to engage with the existing gear $b1$ and the input link. Then, through the combination of teeth, the resulting gear ratios can provide the required rates of the input link. Furthermore, the required rate of the input link can also be generated through the combinations of teeth.

In fact, the identification work of the driving power of the solar subsystem should be regarded as the identification work of the driving power of the whole lost mechanism. The identification results of these arrangements are consistent. Since the lost mechanisms are previously divided into the solar subsystem and the planetary subsystem based on the concept of mechanisms, the driving power arrangements of the lost structures are, respectively, applied to reconstruct their covering subsystems. Therefore, as far as the mechanism with one degree of freedom is concerned, these three single driving power arrangements could be applied to generate different topological structures of mechanisms so as to output the appropriate solar and planetary motions.

8.1.2 Kinematic Analysis of the Solar Theory

The ancient solar theory related to Antikythera device is described in Chap. 1. In the era of Antikythera device, it is well-known that two equivalent systems on geometrical astronomy, the eccentric system and the epicyclic system, exactly demonstrate the solar anomaly motion [12]. Figure 1.9a illustrates these two geometrical systems. As far as the epicyclic system of the solar motion is concerned, its kinematic analysis is the same as the kinematic analysis of the lunar motion. Therefore, the output solar anomaly motion is concluded as expressed in Eq. (7.4). The rates of the epicycle and the deferent (ω_e and ω_d) are based on the description of the solar theory.

Figure 1.9a illustrates the solar motion of the eccentric system. The circle is the real path of the solar motion. The Earth is at an eccentric distance e from the center of the circular path. The distance between the position of the Sun and the center of the path circle O, i.e., the radius of the circle, is constant. However, the distance between the Sun and the Earth is a variable. Based on the definition of the geometrical model, the eccentric system is regarded as a pin-in-slot device, as shown in Fig. 8.2. The rotation center of the link with a pin is the center of the circular path, and the rotation center of the link with a slot is the Earth. The radius of the circular path is much larger than the eccentric distance. Therefore, the kinematic analysis of the eccentric system of the solar motion is the kinematic analysis of the pin-in-slot device. The rate of the link with a slot ω_s is regarded as the output solar motion. In addition, the link with a pin ω_p is regarded as the rotational rate of the Sun about the center of the circular path. In accordance with the concept, the output solar motion is described by Eq. (7.8).

In summary, the kinematic analysis of two systems of the geometrical astronomy is introduced completely. Their concluded output functions are contributed to identify the possible topological structure of mechanisms and the tooth calculation of reconstruction designs.

8.1.3 Eccentric System of the Solar Motion

It is already known that the pin-in-slot joint, which could generate the variable rotational rate, is a critical design of the Antikythera device. The concept has always been emphasized in the reconstruction research of the device. The eccentric system of the solar theory is introduced clearly in Chap. 1. The system is a stationary eccentric. Actually, the simplest design to demonstrate the eccentric system is obtained by direct observation of the system. Based on the concept of mechanisms, it is a 3-bar mechanism with 1 pin-in-slot joint and 2 revolute joints. The link with the pin is the input link, and the link with the slot is the output link. These links, respectively, rotate about points O and E, as shown in Fig. 8.2. The design is without the transformation of rotational direction, i.e., the link with the pin and the link with the slot rotate in the same direction. However, the key is that the simplest design is just a linkage mechanism. It seems to be inconsistent with the surviving evidence, since the Antikythera device is formed by gear mechanisms. Therefore, in order to satisfy the design requirements of gear mechanisms, the feasible arrangements focusing on the design of ordinary gear trains with a pin-in-slot joint should be discussed further.

Based on mobility analysis, the following expression is true for a solar subsystem with 1 degree of freedom, N_L members, N_{JR} revolute joints, N_{JG} gear joints, and N_{JA} pin-in-slot joints:

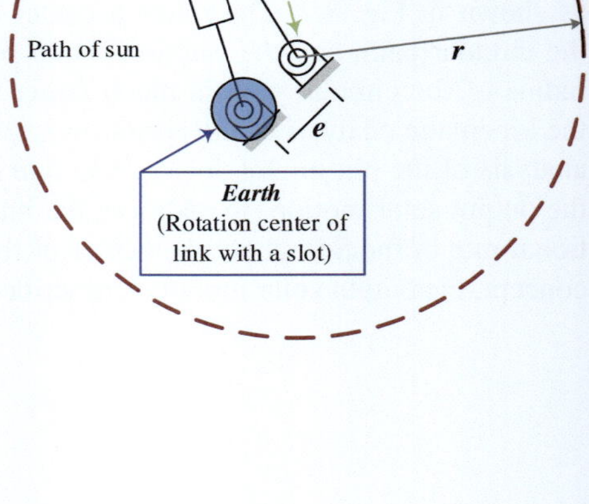

Fig. 8.2 Illustration of a pin-in-slot device and the eccentric system

$$2N_{\mathrm{JR}} + N_{\mathrm{JG}} + N_{\mathrm{JA}} - 3N_{\mathrm{L}} + 4 = 0 \tag{8.1}$$

It is obvious that the following expression is also true for a gear train with N_{J} joints:

$$N_{\mathrm{JR}} + N_{\mathrm{JG}} + N_{\mathrm{JA}} - N_{\mathrm{J}} = 0 \tag{8.2}$$

Since every gear in a gear train must at least have a revolute joint, the following expression is true:

$$N_{\mathrm{JR}} - N_{\mathrm{L}} + 1 = 0 \tag{8.3}$$

It is known that the eccentric system is equivalent to the pin-in-slot device. Since the eccentric system is a stationary eccentric, it is defined for the solar subsystem that N_{JA} is equal to 1. Thus, by solving Eqs. (8.1), (8.2), and (8.3) in terms of the number of links, the following relations for the number of joints are concluded:

$$N_{\mathrm{J}} = 2N_{\mathrm{L}} - 3 \tag{8.4}$$

$$N_{\mathrm{JR}} = N_{\mathrm{L}} - 1 \tag{8.5}$$

$$N_{\mathrm{JG}} = N_{\mathrm{L}} - 3 \tag{8.6}$$

Equations (8.4), (8.5), and (8.6) can calculate the numbers of links and different joints. Furthermore, these equations derive the topological structure of mechanisms with minimum number of links, under the condition of one degree of freedom, i.e., one independent input. For the ordinary gear trains, the number of gear joints, the output rotational direction, and the input rotational direction are closely related to each other. Whatever the rotational directions of the input link is, the actual solar motions must be consistent, i.e., the Sun pointer must display clockwise on the front zodiac dial. Therefore, it is known that the number of gear joints restricts the selection of the driving power arrangements. And, this affects the topological structure of mechanisms of the possible designs of the solar subsystem.

In summary, once the feasible designs of the solar subsystem are ordinary gear trains, the characteristics of the topological structure of mechanisms are obtained by Eqs. (8.4)–(8.6). Furthermore, considering the appropriate driving power arrangements, the topological structures with the minimum number of links lead to two design examples. The first one is that when the solar mechanism is a 4-bar mechanism with 3 revolute joints, 1 gear joint, and 1 pin-in-slot joint, the direction of the input link must rotate counterclockwise. The second one is that when the solar mechanism is a 5-bar mechanism with 4 revolute joints, 2 gear joints, and 1 pin-in-slot joint, the direction of the input link must rotate clockwise.

8.1.4 Epicyclic System of the Solar Motion

Following the research of M.T. Wright, the epicyclic system for the solar motion is demonstrated by the design of an epicyclic gear train. The epicycle is the planet gear, and the deferent is the carrier. The necessary pin-in-slot joint is incident to the planet gear. As mentioned above, the number of gear joints, the direction of the output rotation, and the direction of the input rotation affect each other. The appropriate combinations of the number of gear joints and the input arrangement make the mechanism generate the required output. However, gear joints in an epicyclic gear train, rather than in an ordinary gear trains, can be incident to the frame, i.e., the sun gear can be fixed. Even though the number of gear joints and the rotation direction of the driving power are defined, the known arrangements are not sufficient to determine the output rotational direction. For this reason, the output rotation directions of the epicyclic gear train must be discussed through the identification of links to which gear joints are incident. Afterward, the appropriate driving power arrangement to operate the planetary subsystem can be chosen.

As far as the designs of the epicyclic gear train are concerned, the numbers of links and joints of possible mechanisms can be calculated by Eqs. (8.4), (8.5), and (8.6). According to the design concept with simpler structures, the following two design examples are obtained: The epicyclic gear train is a 4-bar mechanism with 3 revolute joints, 1 gear joint, and 1 pin-in-slot joint; and the epicyclic gear train is a 5-bar mechanism with 4 revolute joints, 2 gear joints, and 1 pin-in-slot joint. In what follows, these two design examples corresponding to various input arrangements are presented in detail.

8.1.4.1 Four-Bar Mechanism with 5 Joints

For the design example of 4-bar mechanism with 3 revolute joints, 1 gear joint, and 1 pin-in-slot joint, the epicyclic gear train includes a sun gear, a carrier, an output gear, and a planet gear at least. Owing to the output gear, the planet gear and the carrier must have revolute joints; and there is no revolute joint incident to the sun gear. Therefore, it is confirmed that the sun gear must be fixed as the ground link. Thus, the carrier is the input link. Furthermore, in order to check whether the designs of epicyclic gear train are able to demonstrate the solar motion, its rate analysis is discussed.

To consider the epicyclic gear train with n planet gears engaging in series ($n = 1, 2, \ldots$), the corresponding rate analysis is discussed as follows:

1. Assume all members of the epicyclic gear train rotate at x rpm. The rates of gears are obtained as follows:

$$\omega'_{\mathrm{s}} = \omega'_{\mathrm{c}} = \omega'_{\mathrm{p}n} = x \tag{8.7}$$

where ω'_{s}, ω'_{c}, and $\omega'_{\mathrm{p}n}$ are the rates of the sun gear, the carrier, and the nth planet gear without considering the engaging of gears, respectively.

2. The carrier does not contribute to the motion of the epicyclic gear train, i.e., assume the sun gear rotates at y rpm, when the carrier is fixed. By the engaging of gears, the rates of gears are obtained as follows:

$$\omega_{\mathrm{s}}'' = y, \quad \omega_{\mathrm{c}}'' = 0, \quad \omega_{\mathrm{p}n}'' = (-1)^n \left(\frac{T_{\mathrm{s}}}{T_{\mathrm{p}n}} \right) y \tag{8.8}$$

where ω_{s}'', ω_{c}'', and $\omega_{\mathrm{p}n}''$ are the rates of the sun gear, the carrier, and the nth planet gear with only considering the engaging of gears, respectively; and T_{s} and $T_{\mathrm{p}n}$ are the teeth of the sun gear and the nth planet gear, respectively.

3. Superimposing the two motions mentioned above, the absolute rates are concluded as follows:

$$\omega_{\mathrm{s}} = x + y, \quad \omega_{\mathrm{c}} = x, \quad \omega_{\mathrm{p}n} = x + (-1)^n \left(\frac{T_{\mathrm{s}}}{T_{\mathrm{p}n}} \right) y \tag{8.9}$$

where ω_{s}, ω_{c}, and $\omega_{\mathrm{p}n}$ are the absolute rates of the sun gear, the carrier, and the nth planet gear, respectively.

In order to simulate the solar anomaly motion, the carrier, i.e., the input link, and the planet gear incident to the pin-in-slot joint are the deferent and the epicycle, respectively. They rotate at the same rate as the tropical year (ω_{year}), but have opposite directions. Therefore, based on the above rate analysis, the following expressions are true:

$$\omega_{\mathrm{c}} = -\omega_{\mathrm{p}n}''$$

or

$$-x = (-1)^n \left(\frac{T_{\mathrm{s}}}{T_{\mathrm{p}n}} \right) y$$

or

$$x = (-1)^{n+1} \left(\frac{T_{\mathrm{s}}}{T_{\mathrm{p}n}} \right) y \tag{8.10}$$

Since the 4-bar mechanism with 5 joints only has 1 planet gear ($n = 1$) and the sun gear must be fixed, the following expressions are true:

$$x = \left(\frac{T_{\mathrm{s}}}{T_{\mathrm{p}n}} \right) y \tag{8.11}$$

$$x + y = 0 \tag{8.12}$$

By solving Eqs. (8.11) and (8.12), the rotation relation yields the following:

$$\left(\frac{T_{\mathrm{s}} + T_{\mathrm{p}n}}{T_{\mathrm{p}n}} \right) y = 0$$

then

$$y = 0 \tag{8.13}$$

According to Eqs. (8.11) and (8.13), x and y are equal to zero. Thus, the epicyclic gear train is motionless. Therefore, the result proves that the design example of the 4-bar epicyclic gear train with 3 revolute joints, 1 gear joint, and 1 pin-in-slot joint is not feasible.

8.1.4.2 Five-Bar Mechanism with 7 Joints

The design example of 5-bar epicyclic gear trains is considered to include a sun gear, a carrier, a planet gear, an output link, and a spare link. The carrier, the planet gear, and the output link must be incident to a revolute joint at least. Then, the assignment of the remaining one revolute joint determines the purpose of the spare link. When the remaining one revolute joint is incident to the spare link, the sun gear must be fixed as the frame; however, when the remaining one revolute joint is incident to the sun gear, the spare link must be fixed as the frame. This situation results in that all members of the epicyclic gear train rotate at the same rate. It erases the function of variable speed by a pin-in-slot joint and does not match the surviving evidence. Therefore, the only feasible way is that the spare link is a transmission link with a revolute joint and the sun gear is fixed.

Except for the above relevant mechanism analysis to discuss the possible topological structure of mechanisms, the kinematic analysis of epicyclic gear train is also one important factor. Focusing on the combination situations of planet gears, the following is the rate analysis of an epicyclic gear train.

All planet gears are adjacent to each other by a gear joint

The epicyclic gear train has 2 planet gears ($n = 2$) engaging in series in this situation. Based on the solar theory and Eq. (8.10) concluded from the kinematic analysis of the epicyclic gear train, the following expression is true:

$$\omega_{\mathrm{c}} = -\omega''_{\mathrm{p}n}$$

or

$$x = -\left(\frac{T_{\mathrm{s}}}{T_{\mathrm{p}n}}\right)y \tag{8.14}$$

By solving Eqs. (8.12) and (8.14), the following expression is true:

$$\left(\frac{T_{\mathrm{p}n} - T_{\mathrm{s}}}{T_{\mathrm{p}n}}\right)y = 0$$

or

$$T_{\mathrm{p}n} = T_{\mathrm{s}} \tag{8.15}$$

Following the result of Eq. (8.15), once the tooth of the sun gear is equal to the tooth of the planet gear with the pin, the topological structure of the mechanism is feasible.

In summary, the design is feasible when the number of the planet gears engaging in series is even (n is even); on the other hand, when the number of the planet gears engaging in series is odd (n is odd), the design is unfeasible.

Two of the planet gears are adjacent to each other by a pin-in-slot joint

The analysis of this case is introduced in the lunar subsystem in Chap. 7. The rate function of the output gear in this case has been concluded. Thus, based on Eq. (7.4), the rate function of the output gear and the ancient solar theory, the following expressions are true:

$$\begin{aligned}\omega_o &= \left\{x + (-1)^n\left(\frac{T_s T_{pk}}{T_{p(k-1)}T_o}\right)y\right\} + F\left\{-(-1)^{n+1}\left(\frac{T_s T_{pk}}{T_{p(k-1)}T_o}\right)y\right\} \\ &= \omega_{year} + F(-\omega_{year})\end{aligned}$$

or

$$x + (-1)^n \frac{T_s T_{pk}}{T_{p(k-1)}T_o} y = (-1)^{n+2} \frac{T_s T_{pk}}{T_{p(k-1)}T_o} y$$

or

$$x = 0 \tag{8.16}$$

Since the sun gear is fixed, x and y are equal to zero, i.e., the mechanism is motionless. Therefore, this topological structure is not feasible. It is confirmed that 2 planet gears must not be adjacent to each other by a pin-in-slot joint.

To conclude all design examples of the eccentric system and the epicyclic system as well as the possible topological structure of mechanism of the design examples, the reconstruction research of the lost solar subsystem is proceeded based on three feasible design types: Type 1 is an ordinary gear train with 4 links and 5 joints, Type 2 is an ordinary gear train with 5 links and 7 joints, and Type 3 is an epicyclic gear train with 5 links and 7 joints. These three design types are mechanisms with 1 degree of freedom. In addition, 3 possible single-input arrangements provide the required input power to operate the solar subsystem.

8.2 Design Process of the Solar Subsystem

The reconstruction work of the solar subsystem is one of special cases. Its research content of building related historical archives is slightly different to the calendrical subsystem and the lunar subsystem, because the corresponding parts of the excavation are lost. It is understood that the historical archives of the solar subsystem includes the research of astronomy and the kinematic and mechanism analyses, in addition to study the existing designs and decoding. According to historical archives, three design types of the solar subsystem are concluded. It is on the basis of the integrated research of ancient astronomy and mechanisms. In what follows, all feasible design concepts, subject to the concluded design constraints and the technique at that time period, are synthesized systematically through step-by-step demonstration of the proposed design procedure.

8.2.1 Type 1 Design of the Solar Subsystem

Type 1 reconstruction design of the solar subsystem is developed depending on the eccentric system of the solar theory. Following the derived characteristics of the topological structure of mechanism and the appropriate input arrangement, the design specifications are concluded as follows:

1. It is an ordinary gear train.
2. The input must rotate counterclockwise.
3. It is a 4-bar mechanism with 5 joints and 1 degree of freedom.
4. The joints include 3 revolute joints, 1 external gear joint, and 1 pin-in-slot joint.
5. The types of mechanical members are gears and links.

Based on the concepts of generalization and number synthesis of generalized chains, only 1 generalized chain of the 4-bar mechanism with 5 joints is available, as shown in Fig. 8.3.

Through the study of historical literatures as well as the analysis of the eccentric system and the existing design, the design constraints are concluded as follows:

1. To compare with the existing design and the decoding about the front dials and pointers, the output link must rotate around axis *b* that represents the daily motion.
2. According to the kinematical analysis of Hipparchus' solar theory, the solar subsystem must have the function of variable rate.
3. The solar subsystem must have at least a ground link (link 1, K_F), an input link (link 2, K_I), an output link (link 3, K_O), and the transmission link (link 4, K_{T1}).

   Ground link (member 1, K_F)

   a. There must be a ground link as the frame.
   b. For the simple gear train, each remaining link must be adjacent to the ground link; therefore, the ground link must be a multiple link.

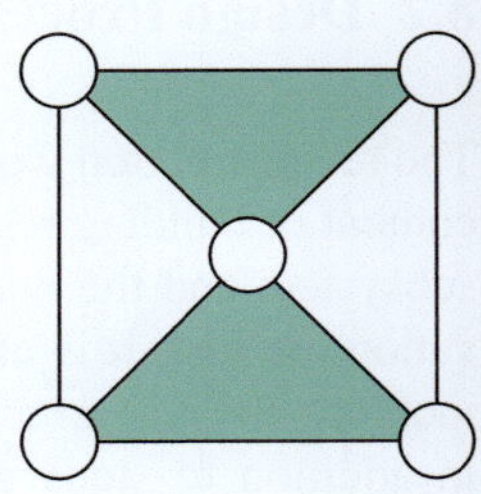

Fig. 8.3 Generalized chain with 4 links and 5 joints

Input link (member 2, K_I)

a. There must be a binary link as the input link.
b. The input link must be adjacent to the ground link.
c. The input link must not be adjacent to the output link to avoid the degeneration of the mechanism.

Output link (member 3, K_O)

a. There must be a binary link as the output link.
b. The input link must be adjacent to the ground link.

Transmission link (member 4, K_T)

a. There must be at least one link as the transmission link.
b. The transmission link must be a multiple link including at least 1 revolute joint and 1 gear joint.
c. The transmission link must be adjacent to the ground link.

Pin-in-slot joint (joint e, J_A)

a. There must be 1 pin-in-slot joint, as a cam joint, to generate the function of variable rate.
b. The two links connected to the pin-in-slot joint must not rotate around the common axis.

Revolute joint (joints a, b, c; J_R)

a. There must be 3 revolute joints.
b. Every link must have at least 1 revolute joint.
c. Any joint incident to the ground link must be a revolute joint.
d. There can be no loop formed exclusively by revolute joints.

Gear joint (joint d, J_G)

a. There can be no 3-bar loop formed exclusively by gear joints.
b. Owing to the input member being in counterclockwise rotation, opposite to the display of the solar motion, the number of gear joints must be odd, i.e., there must be a gear joint.

Then, according to the design constraints concluded above, all specialized chains of Type 1 design of the solar subsystem are identified by the following steps:

Step 1. For the available generalized chain, the ground link is assigned.
Step 2. For each result from Step 1, the input link is assigned.
Step 3. For each result from Step 2, the output link is assigned.
Step 4. For each result from Step 3, the remaining links are assigned as the transmission links.
Step 5. For each result from Step 4, the pin-in-slot joint is assigned.
Step 6. For each result from Step 5, the revolute joints are assigned.
Step 7. For each result from Step 6, the gear joint is assigned.

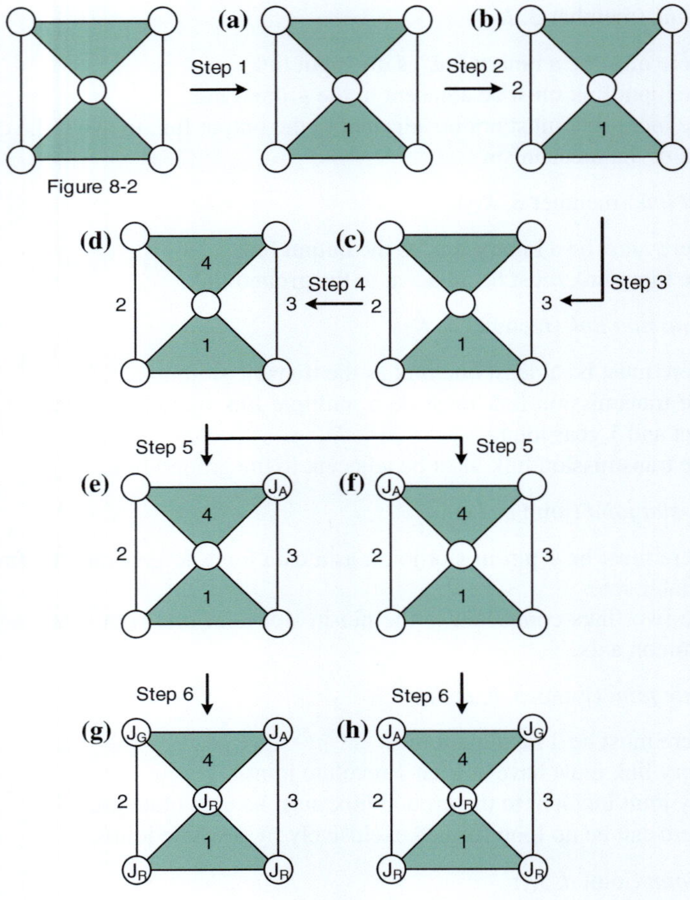

Fig. 8.4 Specialization process of Type 1 design of the solar subsystem

For the only one generalized chain with 4 links and 5 joints shown in Fig. 8.3, its specialization process are illustrated as follows:

1. The ground link must be a ternary link. Since the ternary links of the generalized chain shown in Fig. 8.3 are symmetric, only one of them is identified as the ground link. Therefore, 1 specialized chain with identified ground link is available, as shown in Fig. 8.4a.
2. The input link must be a binary link, which is adjacent to the ground link. Based on the concept of similar classes, only one of the two binary links of the specialized chain with identified ground link shown in Fig. 8.4a can be identified as input link. Therefore, 1 specialized chain with identified ground link and input link is available, as shown in Fig. 8.4b.
3. The output link must be a binary link adjacent to the ground link. Therefore, 1 specialized chain with identified ground link, input link, and output link is available, as shown in Fig. 8.4c.

4. The remaining link is identified as the transmission link. Therefore, 1 specialized chain with identified ground link, input link, output link, and transmission link is available, as shown in Fig. 8.4d.
5. The pin-in-slot joint must not be incident to the ground link. For the specialized chain with identified ground link, input link, output link, and transmission link shown in Fig. 8.4d, the assignment of the pin-in-slot joint generates 2 results, as shown in Fig. 8.4e, f.
6. There must be 3 revolute joints in each design. Type 1 design is an ordinary gear train. Therefore, these three revolute joints are incident to the ground link, as shown in Fig. 8.4g, h.
7. Since there must be a gear joint in each design, the remaining joint is identified as a gear joint, as shown in Fig. 8.4g, h.

Therefore, 2 specialized chains with identified ground link, input link, output link, transmission link, pin-in-slot joint, revolute joints, and gear joint are available, as shown in Fig. 8.4g, h. Furthermore, Type 1 reconstruction design of the solar subsystem has one gear joint to change the direction of rotation. The design specifications indicate that the input link must rotate once every year. Therefore, based on the investigation of the possible input arrangements with a single input, only input arrangement 2 shown in Fig. 8.1b is satisfied. Thus, the input link (member 2) has to engage with the existing contrate gear $a1$.

It is known that the pointers of the Sun and the date must rotate coaxially in the geometrical constraint of mechanism. Two specialized chains shown in Fig. 8.4g, h are feasible, and they are particularized into corresponding kinematic drawings, as shown in Fig. 8.5a, b. Therefore, two feasible reconstruction designs of the solar

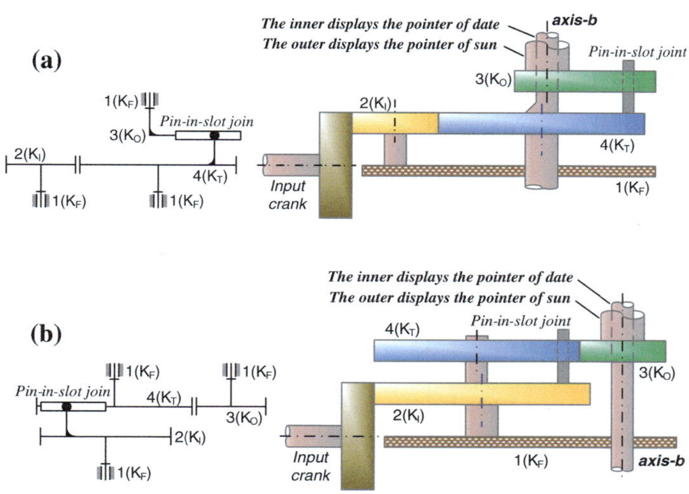

Fig. 8.5 Type 1 feasible designs of the solar subsystem

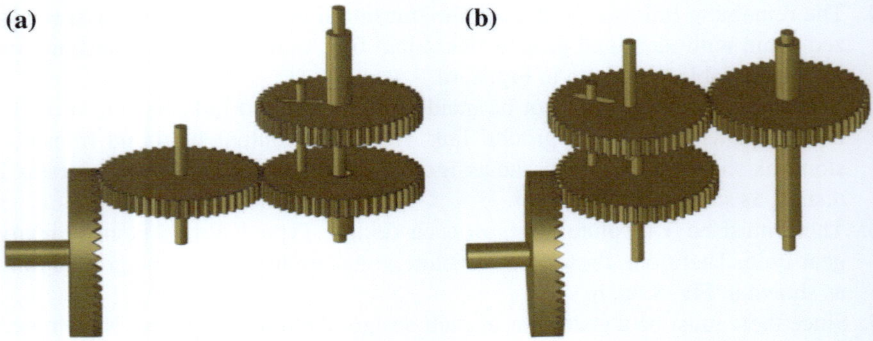

Fig. 8.6 3D models of Type 1 feasible designs. **a** Corresponding to Fig. 8.5a, **b** corresponding to Fig. 8.5b

subsystem are obtained. Figure 8.6 shows 3D solid models of the feasible reconstruction designs.

Furthermore, based on the known input arrangement, the required output characteristic (anomaly solar motion), and the kinematic analysis of gear train, the relations of tooth calculation of these two feasible reconstruction designs are concluded as follows:

1. Feasible reconstruction design shown in Fig. 8.5a:
Based on Eq. (7.8), the relation between the rates of the output link (member 3) and the transmission link (member 4) is derived as follows:

$$\omega_3 = A\omega_4 \tag{8.17}$$

where A is the anomaly function mentioned in the kinematic analysis of the solar theory in Chap. 1. In addition, since the transmission link (member 4) has to rotate at the rate of the tropical year, which equals to the rate of the surviving gear $b1$, the following expression is true:

$$\frac{\omega_3}{\omega_{b1}} = \frac{A\omega_4}{\omega_{b1}} = A\frac{T_{b1} \times T_{a1} \times T_2}{T_{a1} \times T_2 \times T_4} = A\frac{T_{b1}}{T_4}$$

or

$$T_4 = T_{b1} = 224 \tag{8.18}$$

2. Feasible reconstruction design shown in Fig. 8.5b:
Similarly, based on Eq. (7.8), the relation between the rates of transmission link (member 4) and the input link (member 2) is derived as:

$$\omega_4 = A\omega_2 \tag{8.19}$$

Then, according to the relations of the gear ratios, the following expressions are true:

$$\omega_3 = \omega_4 \left(\frac{T_4}{T_3} \right) \tag{8.20}$$

$$\frac{\omega_3}{\omega_{b1}} = \frac{(T_4/T_3)\omega_4}{\omega_{b1}} = \frac{A(T_4/T_3)\omega_2}{\omega_{b1}} = A\frac{T_{b1} \times T_4}{T_2 \times T_3}$$

or

$$\frac{T_{b1} \times T_4}{T_2 \times T_3} = 1 \tag{8.21}$$

8.2.2 Type 2 Design of the Solar Subsystem

Like Type 1 reconstruction design, Type 2 reconstruction design of the solar subsystem is also developed from Apollonius' and Hipparchus' eccentric system. Based on the analysis of the feasible topological structure of mechanisms mentioned above, this type should be the design of an ordinary gear train. Moreover, it is identified that the input link has to rotate clockwise. As a result, the corresponding design specifications are concluded as follows:

1. It is an ordinary gear train.
2. The input link must rotate clockwise.
3. It is a 5-bar mechanism with 7 joints and 1 degree of freedom.
4. The joints include 4 revolute joints, 2 external gear joints and 1 pin-in-slot joint.
5. The types of mechanical members are gears and links.

According to the concepts of generalization and number synthesis of generalized chains, the atlas of generalized chains with 5 links and 7 joints is generated, as shown in Fig. 8.7. Next, the design constraints has to be considered and concluded. Since Type 1 and Type 2 reconstruction designs depend on the eccentric system, most design constraints of these two types are the same. However, some

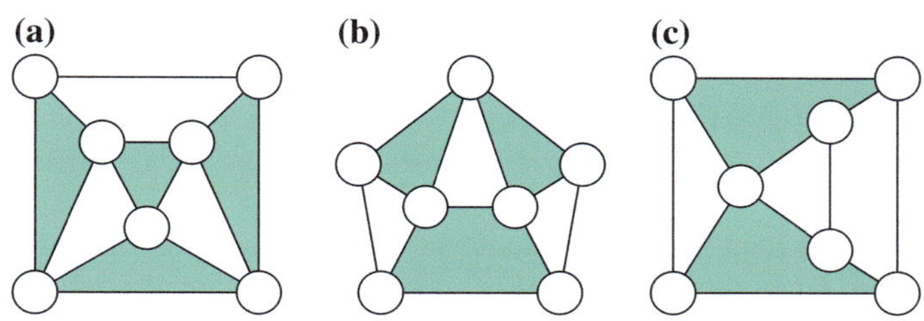

Fig. 8.7 Atlas of generalized chains with 5 links and 7 joints

design constraints that are different from Type 1 reconstruction design are concluded as follows:

1. The solar subsystem must have a ground link (member 1, K_F), an input link (member 2, K_I), an output link (member 3, K_O), and the transmission links (member 4, K_{T1}; member 5, K_{T2}).

   Transmission links (member 4, K_{T1}; member 5, K_{T2})

   a. There must be at least 2 links as the transmission links.
   b. The transmission links must be a multiple link including at least 1 revolute joint and 1 gear joint.

   Revolute joint (joints *a*, *b*, *c*, *d*; J_R)

   a. In accordance with the mechanism analysis derived above, there must be 4 revolute joints.

   Gear joint (joints *e* and *f*, J_G)

   a. Owing to the input and the output members being clockwise rotation, the numbers of gear joints must be even, i.e., there must have 2 gear joints.

The specialization process of Type 2 design can be proceeded through the same steps of the link and joint assignments mentioned above. Due to the design constraints that the ground link must have one multiple link and be adjacent to two binary links at least, only the two generalized chains shown in Fig. 8.7b, c are qualified to identify the ground link. Furthermore, there are two transmission links in each design. The generalized chains must have two ternary links at least. Therefore, only one generalized chain shown in Fig. 8.7b is qualified to identify the two transmission links. For the generalized chain shown in Fig. 8.7b, its corresponding specialized chains are identified as follows:

1. Since the ground link must be a multiple link, the assignment of the ground link generates 1 result as shown in Fig. 8.8a.
2. The input link must be a binary link adjacent to the ground link. Therefore, for the specialized chain shown in Fig. 8.8a, based on the concept of similar classes, the assignment of the input link generates 1 result, as shown in Fig. 8.8.
3. The output link must be a binary link adjacent to the ground link. Therefore, for the specialized chain shown in Fig. 8.8b, the assignment of the output link generates 1 result, as shown in Fig. 8.8c.
4. The transmission links must be ternary links adjacent to the ground link. Based on the concept of similar classes, the transmission links are identified, as shown in Fig. 8.8d.
5. The pin-in-slot joint must not be incident to the ground link. Therefore, for the specialized chain shown in Fig. 8.8d, the assignment of the pin-in-slot joint generates 3 results, as shown in Fig. 8.8e–g.

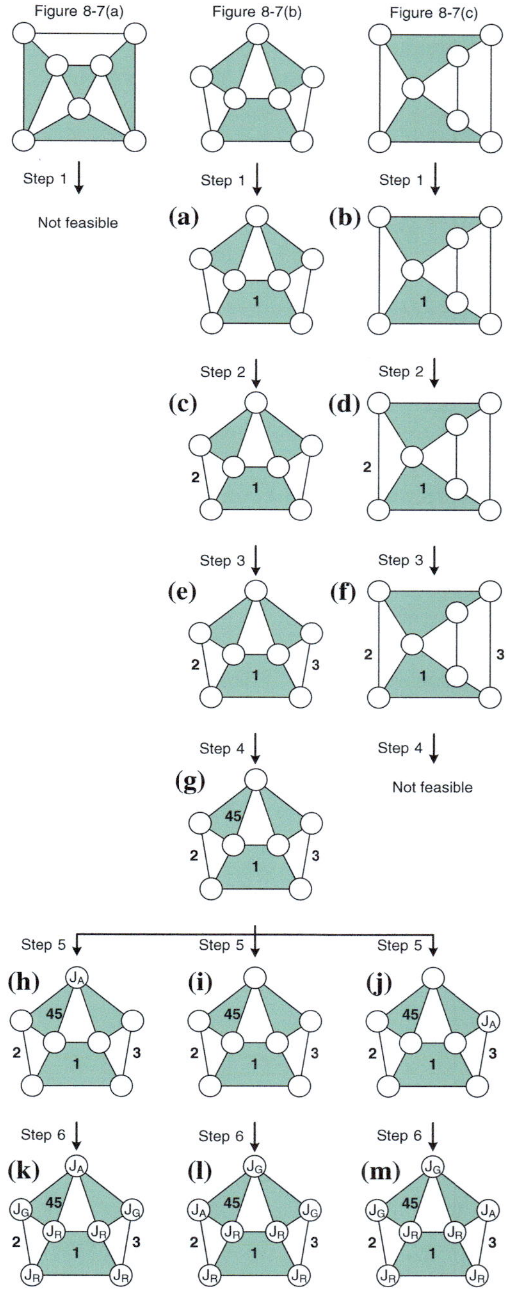

Fig. 8.8 Specialization process of Type 2 design of the solar subsystem

6. All joints incident to the ground link are revolute joints. And, there must be 4 revolute joints in each design. Therefore, the revolute joints are identified from the specialized chains, as shown in Fig. 8.8h–j.
7. Since there must be two gear joints in each design, the remaining joints are identified as gear joints, as shown in Fig. 8.8h–j.

By the design procedure, 3 specialized chains of Type 2 reconstruction design are generated, as shown in Fig. 8.8a–c. However, not all of these 3 specialized chains are feasible. In order to particularize the specialized chains into the mechanical or schematic drawings, the appropriate input arrangement and the geometrical constraints of the mechanism should be examined again. In accordance with the design specification that the input link must rotate clockwise and the concluded input arrangements shown in Fig. 8.1, only the input arrangement shown in Fig. 8.1a can be applied, i.e., the input link with the clockwise rotation is connected to axis *b*, which operates the pointer of date. Under this input arrangement, the input link and the output link must be coaxial. Thus, it is a reverted ordinary gear train. Therefore, considering the above geometrical constraint, the specialized chains shown in Fig. 8.8b, c are not feasible. Only the one shown in Fig. 8.8a is feasible to particularize into the corresponding kinematic drawing, as shown in Fig. 8.9. Figure 8.10 shows the corresponding 3D solid model of this feasible reconstruction design.

After obtaining the feasible reconstruction design, the tooth counting work can be discussed further. Focusing on this feasible design shown in Fig. 8.9, the rate relations between the two transmission links (members 4 and 5), the input link (member 2) and the transmission link (member 4), and the output link (member 3) and the transmission link (member 5), respectively, are concluded by Eq. (7.4) and the gear ratios are as follows:

$$\omega_5 = A\omega_4 \tag{8.22}$$

$$\omega_2 = \left(\frac{T_4}{T_2}\right)\omega_4 \tag{8.23}$$

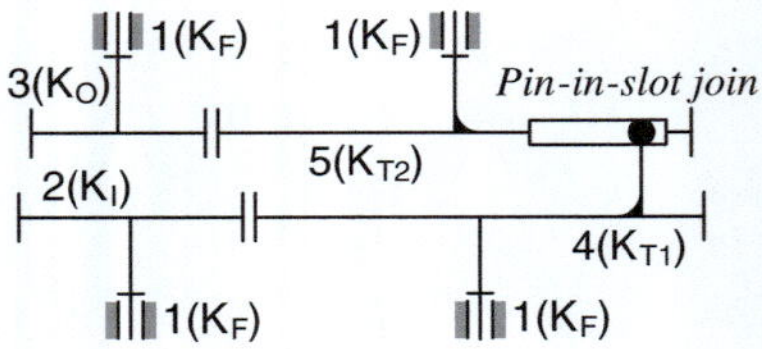

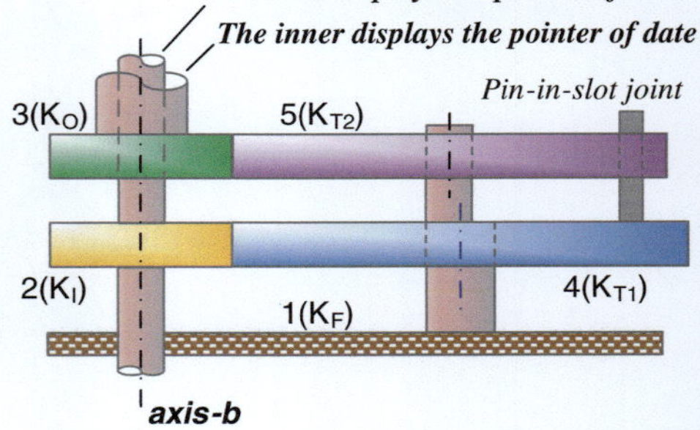

Fig. 8.9 Type 2 feasible design of the solar subsystem

Fig. 8.10 3D model of Type 2 feasible design

$$\omega_3 = \left(\frac{T_5}{T_3}\right)\omega_5 \tag{8.24}$$

Furthermore, all teeth included in the design of Fig. 8.9 must satisfy the following equation:

$$\frac{\omega_5}{\omega_2} = A\frac{T_2 \times T_5}{T_4 \times T_3}$$

or

$$\frac{T_2 \times T_5}{T_4 \times T_3} = 1 \tag{8.25}$$

In summary, the teeth of the reconstruction design of Fig. 8.9 can be appropriately chosen for the detail designs so as to build the corresponding simulation model.

8.2.3 Type 3 Design of the Solar Subsystem

According to the concluded mechanism analysis, Type 3 reconstruction design of the solar subsystem depends on the epicyclic system. This type should be the design of an epicyclic gear train. Since the links incident to the pin-in-slot joint is unknown before the assignments of links and joints, Type 3 reconstruction design is unlike to be an ordinary gear train of previous design types that the output

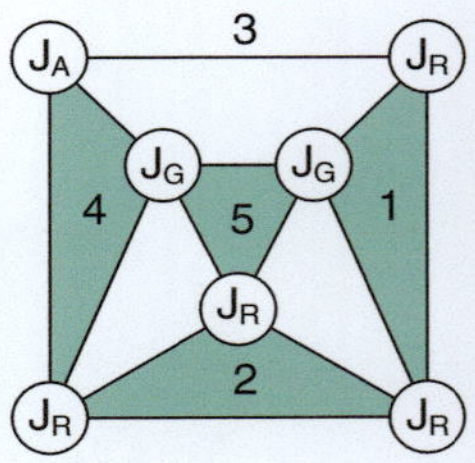

Fig. 8.11 Generalized chain corresponding to Wright's design for the solar motion

rotation direction can be identified based on the input rotation direction and the number of joints. The rotation direction of the input link in this case is uncertain. Therefore, the design specifications of Type 3 reconstruction design of the solar subsystem are concluded as follows:

1. It is an epicyclic gear train.
2. It is a 5-bar mechanism with 7 joints and 1 degree of freedom.
3. The joints include 4 revolute joints, 2 external gear joints, and 1 pin-in-slot joint.
4. The types of mechanical members are gears and links.

According to the concept of generalization, the design by Wright shown in Fig. 4.4 is generalized into its generalized chain, as shown in Fig. 8.11. Furthermore, based on the number synthesis of generalized chains, the atlas of generalized chains with 5 links and 7 joint is obtained, as shown in Fig. 8.7.

Following the previous study of the historical archives, the design constraints of Type 3 reconstruction design are concluded as follows:

1. Based on the decoding of the front dial, it is indispensable that the output link displaying the solar motion and the link displaying the daily motion are coaxial, i.e., the output link rotates around axis *b*.
2. The solar subsystem must have at least a ground link (member 1), an input link (member 2), an output link (member 3), and the transmission links (members 4 and 5).

   Ground link (member 1, K_F)

   a. There must be a ground link as the frame.
   b. The ground link must be a multiple link, including 2 revolute joints and 1 gear joint at least.

   Input link (member 2, K_I)

   a. There must be a link as the input link functioning as the carrier.
   b. The input link must be adjacent to the ground link with a revolute joint.
   c. The input link must not be adjacent to the output link to avoid the degeneration of the mechanism.

   Output link (member 3, K_O)

   a. There must be a binary link as the output link.

Transmission links (member 4, K_{T1}; member 5, K_{T2})

a. There must have two links as the transmission links functioning as planet gears.
b. The transmission links must be multiple links including at least 1 revolute joint and 1 gear joint.
c. The transmission links must not be adjacent to the ground link with revolute joints.
d. The transmission links must be adjacent to the input link with revolute joints.

Pin-in-slot joint (joint *f*, J_A)

a. There must be a pin-in-slot joint to generate variable rate function.
b. A pin-in-slot joint must be adjacent to the transmission link.
c. Two links incident to the pin-in-slot joint must not rotate around the common axis, since of the effect of eccentric.

Revolute joint (joints *a*, *b*, *d*, *e*; J_R)

a. There must have 4 revolute joints.
b. Every link must have at least 1 revolute joint.
c. There can be no loop formed exclusively by revolute joints.

Gear joint (joints *c* and *g*, J_G)

a. There can be no 3-bar loop formed exclusively by gear joints.
b. There must be 1 gear train incident to the ground link at least.

All possible specialized chains corresponding to the generalized chains shown in Fig. 8.7 can be identified based on the assignment procedure mentioned above. According to the concluded design constraints, the specialization process is as follows.

8.2.3.1 *Ground Link* (Member 1)

1. For the generalized chain shown in Fig. 8.7a, the assignment of the ground link generates 2 results, as shown in Fig. 8.12a, b.
2. For the generalized chain shown in Fig. 8.7b, the assignment of the ground link generates 2 results, as shown in Fig. 8.12c, d.
3. For the generalized chain shown in Fig. 8.7c, the assignment of the ground link generates 1 result, as shown in Fig. 8.12e.

Therefore, 5 specialized chains with identified ground link are available, as shown in Fig. 8.12a–e.

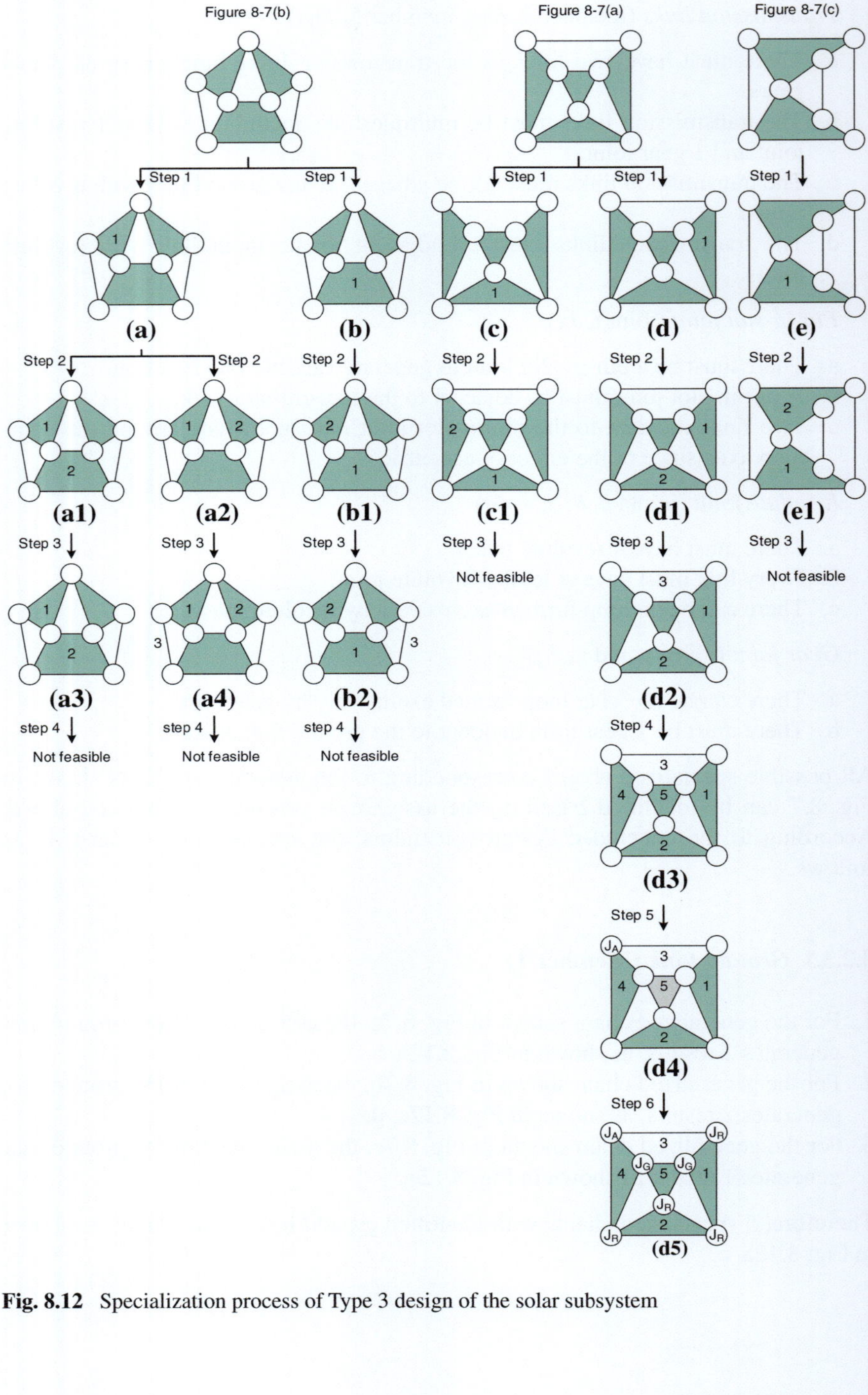

Fig. 8.12 Specialization process of Type 3 design of the solar subsystem

8.2.3.2 *Input Link* (Member 2)

1. For the specialized chain shown in Fig. 8.12a, the assignment of the input link generates 1 result, as shown in Fig. 8.12f.
2. For the specialized chain shown in Fig. 8.12b, the assignment of the input link generates 1 result, as shown in Fig. 8.12g.
3. For the generalized chain shown in Fig. 8.12c, the assignment of the input link generates 2 results, as shown in Fig. 8.12h, i.
4. For the generalized chain shown in Fig. 8.12d, the assignment of the input link generates 1 result, as shown in Fig. 8.12j.
5. For the generalized chain shown in Fig. 8.12e, the assignment of the input link generates 1 result, as shown in Fig. 8.12k.

Therefore, 6 specialized chains with identified ground link and input link are available, as shown in Fig. 8.12f–k.

8.2.3.3 *Output Link* (Member 3)

1. For the specialized chain shown in Fig. 8.12f, the assignment of the output link generates zero results.
2. For the specialized chain shown in Fig. 8.12g, the assignment of the output link generates 1 result, as shown in Fig. 8.12l.
3. For the generalized chain shown in Fig. 8.12h, the assignment of the output link generates 1 result, as shown in Fig. 8.12m.
4. For the generalized chain shown in Fig. 8.12i, the assignment of the output link generates 1 result, as shown in Fig. 8.12n.
5. For the generalized chain shown in Fig. 8.12j, the assignment of the ground link generates 1 result, as shown in Fig. 8.12o.
6. For the generalized chain shown in Fig. 8.12k, the assignment of the ground link generates zero results.

Therefore, 4 specialized chains with identified ground link, input link, and output link are available, as shown in Fig. 8.12l–o.

8.2.3.4 *Transmission Links* (Members 4 and 5)

All transmission links are ternary links. For the specialized chains with identified ground link, input link, and output link shown in Fig. 8.12l–o, their remaining links must be ternary links so as to be taken as the transmission links. Therefore, only the specialized chain with identified ground link, input link, and output link shown in Fig. 8.12l is qualified to have the transmission links, as shown in Fig. 8.12p.

8.2.3.5 *Pin-in-Slot Joint* (Joint J_A)

The pin-in-slot joint must not be incident to the ground link. Since Type 3 design is to simulate the epicyclic system of the solar motion, the pin-in-slot joint must be incident to the output link. Therefore, the pin-in-slot joint is identified, as shown in Fig. 8.12q.

8.2.3.6 *Revolute Joints* (Joint J_R)

There are 4 revolute joints in each design. Each link of the designs must have one revolute joint at least. The ground link must be a fixed gear, and the input link must be a carrier. Therefore, the revolute joints are identified, as shown in Fig. 8.12r.

8.2.3.7 *Gear Joints* (Joint J_G)

There are 2 gear joints in each design. Therefore, the remaining joints are identified as gear joints, as shown in Fig. 8.12r.

Only one specialized chain is generated by illustrating the design procedure. This specialized chain can be particularized into the corresponding mechanical and schematic drawing, as shown in Fig. 8.13. The design of the epicyclic gear train, shown in Fig. 8.13, can generate various reconstruction designs of the solar subsystem by choosing different driving arrangements. Therefore, the number of the feasible reconstruction designs is determined after identifying the appropriate driving arrangements. In accordance with the kinematic analysis mentioned in the historical archives, the input link (member 2; carrier) must rotate clockwise at the rate of a tropical year. As a result, the known rotational direction and the rate of the input link are contributed to choose the exact driving modes from the single-input driving arrangements, as shown in Fig. 8.1.

According to the rotational direction, two single-input driving arrangements shown in Fig. 8.1a, c are satisfied. The combinations of the reconstruction design and these two driving arrangements are shown in Fig. 8.14. Furthermore,

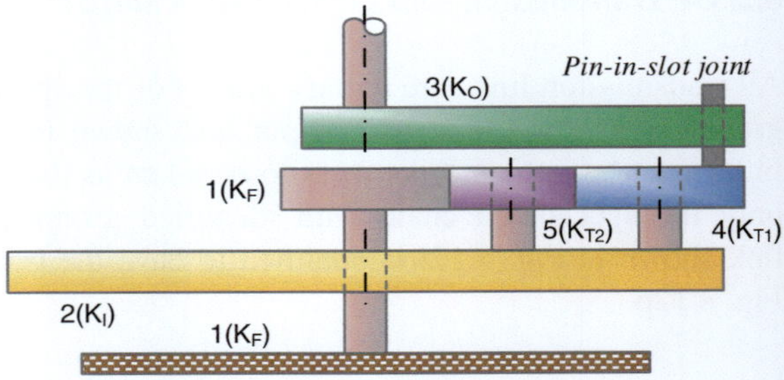

Fig. 8.13 Mechanical drawing of the feasible specialized chain of Type 3 design

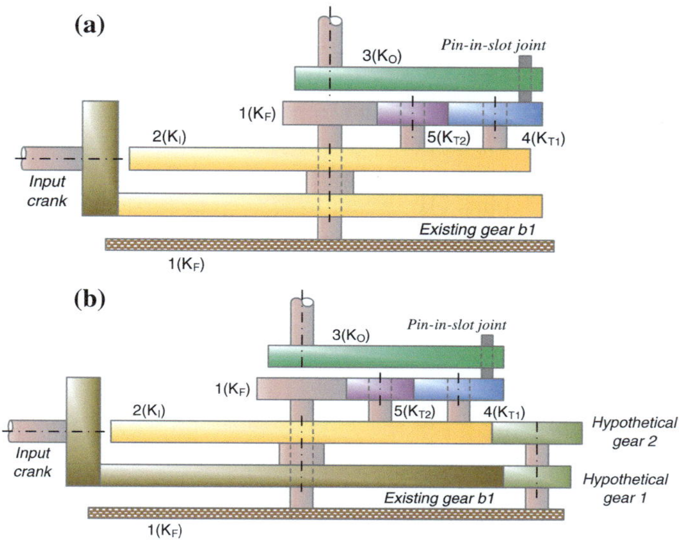

Fig. 8.14 Type 3 feasible designs of the solar subsystem

Fig. 8.15 3D model of Type 3 feasible design

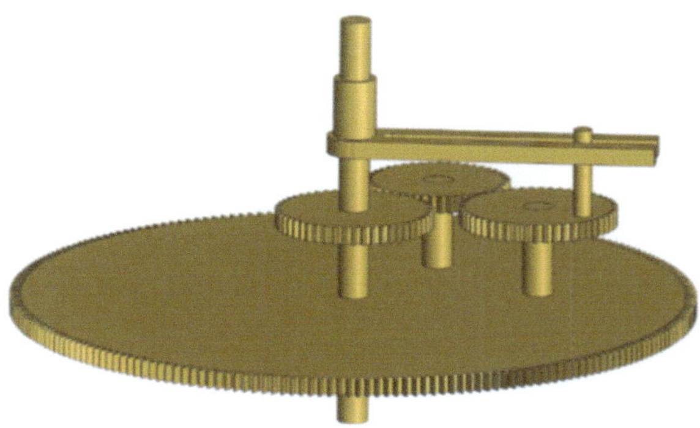

by considering the rotational rate, the reconstruction design shown in Fig. 8.14a results in the same teeth of the hypothetical compound gear in order to generate the rate of a tropical year. It makes the hypothetical compound gear of the design redundant. For this reason, the reconstruction design shown in Fig. 8.14b is more appropriate, i.e., the input link of the reconstruction design is driven by mechanically connecting to axis *b*. Also, it is the existing design by Wright. Figure 8.15 shows the 3D solid model of the feasible reconstruction design of Type 3 design. Finally, based on Eq. (8.15), the relation of teeth is derived as follows:

$$T_4 = T_1 \tag{8.26}$$

The teeth of the input link (member 2) and the transmission link (member 5) carrier are not constrained. They are designed appropriately to satisfy the geometrical constraints of the interior mechanism of Antikythera device.

In summary, the reconstruction designs of the solar subsystem are proceeded by three design types: Type 1 is an ordinary gear train with 4 links and 5 joints, Type 2 is an ordinary gear train with 5 links and 7 joints, and Type 3 is an epicyclic gear train with 5 links and 7 joints. According to the design procedure, the concluded design constraints, and the possible single-input driving arrangements, these three types generate, respectively, 2, 1, and 1 feasible reconstruction designs. Type 3 reconstruction design is the existing design, and the remaining two reconstruction designs are alternative solutions.

8.3 Remarks

The lost solar subsystem is reconstructed, and 4 feasible designs, respectively, from three design types are generated. Nevertheless, the lost excavation is still a key. Only the design types with fewer elements are applied in this chapter, since a simple structure is better for achieving the purpose of a mechanical device, from mechanical designers' viewpoint. In the future, once new evidences are discovered and the adding of element numbers is essential, the design methodology presented in Chap. 5 is still workable to synthesize all feasible reconstruction designs.

References

1. Lin JL, Yan HS (2014) Historical development of reconstruction designs of Antikythera mechanism. In: Applied mechanics and materials, vol 163, pp 1–6, Apr 2012. Presented in proceedings of the 9th international conference on history of mechanical technology and mechanical design, Tainan, 23–25 Mar 2014
2. Yan HS (2007) Reconstruction designs of lost ancient Chinese machinery. Springer, Netherland
3. Yan HS (1992) A methodology for creative mechanism design. Mech Mach Theory 27(3):235–242
4. Yan HS (1998) Creative design of mechanical devices. Springer, Singapore
5. Yan HS, Lin JL (2013) Reconstruction synthesis of the lost interior structure for the solar anomaly motion of the Antikythera mechanism. Mech Mach Theory 70:354–371
6. Evans J, Carman CC, Thorndike AS (2010) Solar anomaly and planetary displays in the Antikythera mechanism. J Hist Astron 42:1–39
7. Carman CC, Thorndike AS, Evans J (2012) On the pin-and-slot device of the Antikythera mechanism, with a new application to the superior planets. J Hist Astron 43(150):93–116
8. Evans J, Carman CC (2014) Mechanical astronomy: a route to the ancient discovery of epicycles and eccentrics. In: Sidoli N, Van Brummelen G (eds) Alexandria, through Baghdad: surveys and studies in the ancient greek and medieval islamic mathematical sciences in honor of J.L. Berggren. Springer, Berlin, pp 145–171
9. Wright MT (2002) A planetarium display for the antikythera mechanism. Horological J 144(5):169–173

10. Wright MT (2003) Epicyclic gearing and the Antikythera mechanism: part 1. Antiquarian Horology 27:270–279
11. Freeth T, Jones A (2012) The cosmos in the Antikythera mechanism. In: Institute for the study of the ancient world. New York University. Available in the website: Institute for the Study of the Ancient World (ISAW). http://dlib.nyu.edu/awdl/isaw/isaw-papers/4/. Accessed July 2014
12. Linton CM (2004) From Eudoxus to Einstein: a history of mathematical astronomy. Cambridge University, New York

Chapter 9
Reconstruction Designs of the Planetary Subsystem

The planetary subsystem of Antikythera device can demonstrate completely the planetary pointers on the front zodiac dial, including Mercury, Venus, Mars, Jupiter, and Saturn. Several studies, based on the concept of mechanism designs, presented reconstruction models with the interior mechanisms serving to the planetary functions [1–5]. However, the surviving evidence of planetary subsystem and the solar subsystem is not complete. Even some designs have been presented, a systematic discussion regarding alternative feasible topological structures of mechanisms of the lost planetary subsystem is needed.

In this chapter, the systematical design procedure presented in Chap. 5 is applied again to reconstruct all feasible designs of the planetary subsystem [6–10]. The driving power arrangements of the lost planetary subsystem are analyzed. And, the relations of tooth calculation for each feasible design are concluded

9.1 Historical Archives of the Planetary Subsystem

The historical archives of the planetary subsystem are built based on the identifications of the possible driving arrangements and the integrated study of ancient astronomy and mechanisms. The integrated study includes the kinematic analysis of planetary theory and gear trains with a pin-in-slot and the mechanism analysis to conclude the numbers of links and joints. In fact, the related researches to build the historical archives are almost the same as the process of the solar subsystem. Most works of the historical archives of the planetary subsystem are presented in Chap. 4. Three concluded single driving power arrangements are applied to operate the planetary subsystem, as shown in Fig. 8.1. The motion equations to simulate the planetary motions are concluded by the kinematic analysis of gear trains with a pin-in-slot joint. The remaining work of the historical archives is to determine the possible topological structures, including the types and the corresponding

J.-L. Lin and H.-S. Yan, *Decoding the Mechanisms of Antikythera Astronomical Device*, DOI 10.1007/978-3-662-48447-0_9

numbers of links and joints of the mechanism. And, this work has to be achieved by the kinematic analysis of ancient planetary theory.

The fact that Apollonius' planetary theory existed in the era of Antikythera device has been introduced clearly in Chap. 1. Figure 1.9b shows the epicyclic subsystem of the planetary motions. Its kinematic analysis is similar to that of the solar theory in Chap. 8. The difference is the rotational direction. Therefore, the motion equation to describe the planetary motions based on the planetary theory is expressed in Eq. (7.4). Absolutely, all of the cyclic motions of the five planets known in the ancient time can be depicted by the geometrical system. In addition, their motions are simulated completely by the concluded Eq. (7.4).

The kinematic analysis of the related ancient astronomical theories is contributed to conclude the mechanism types of the unknown planetary subsystem. The motion equation, Eq. (7.4), concluded from the epicyclic system of the planetary motions displays the planetary anomaly motions. The output rate is the rate of the deferent plus the anomaly function affected by the rate difference between the deferent's rate and the epicycle's rate. Therefore, for the designs of gear trains, epicyclic gear trains are the only feasible type of gear trains. The possible topological structures of the planetary subsystem should be analyzed further.

Through the mechanism analysis mentioned in the reconstruction designs of the solar subsystem, the relations between the numbers of joints and links are derived, as expressed in Eqs. (8.4), (8.5), and (8.6). The concluded single driving power arrangements of the planetary subsystem are with clockwise or counterclockwise direction. The planetary subsystem must output the clockwise rotation on the front dial. Therefore, the gear train for the planetary subsystem should have at least one gear joint to change the rotation direction.

Thus, based on the number of gear joints and the design concept of simple structure, two design types of the planetary subsystem are presented: Type 1 is an epicyclic gear train with one gear joint, and Type 2 is an epicyclic gear train with two gear joints. In what follows, these two types are presented in detail.

9.1.1 Type 1 Design: Mechanism with One Gear Joint

For the planetary subsystem with only one gear joint, based on Eqs. (8.4)–(8.6), this mechanism has 4 links, 3 revolute joints, 1 gear joint, and 1 pin-in-slot joint. Furthermore, the mechanical members of the epicyclic train include a sun gear, a carrier, a planet gear, and an output link (gear) at least. The output link (gear), the planet gear, and the carrier must have revolute joints to transmit the rotational motion. These three mechanical elements require three revolute joints. By inspecting the numbers of gear joints of this type, there is no revolute joint adjacent to the sun gear. Therefore, the sun gear must be fixed as the ground link.

9.1.2 Type 2 Design: Mechanism with Two Gear Joints

For this type, the mechanism has 5 links, 4 revolute joints, 2 gear joints, and 1 pin-in-slot joint, based on Eqs. (8.4), (8.6). Since the necessary members of an epicyclic gear train include a sun gear, a carrier, a planet gear, and an output link (gear), the role of the remaining member of this epicyclic gear train should be clarified further. Through the same discussion process mentioned in the solar subsystem, i.e., the assignment of the remaining revolute joint, it is confirmed that the remaining element of the epicyclic gear train must play as a planet gear like a transmission link with a revolute joint. Therefore, the sun gear must be fixed as the frame.

Kinematic analysis of epicyclic gear trains is helpful to conclude the characteristics of the topological structures. In addition, the outputs of the epicyclic gear train should be inspected if it satisfies the Apollonius' planetary theory. The analysis process is also the same as the process of the solar subsystem. Therefore, to imitate the analysis process of the solar subsystem, the following two design conditions are presented.

9.1.2.1 All Planet Gears Are Adjacent to Each Other by a Gear Joint

By the inspection of the Apollonius' planetary theory and the hypothetical condition of an epicyclic gear train, the planet gears incident to a pin-in-slot joint and the carrier are regarded as the epicycle and the deferent, respectively. They must rotate in the same direction and with the corresponding rates. Therefore, according to the results of the kinematic analysis mentioned in Chap. 7, it is concluded that when the number of the planet gears engaging in series is even (n is even), the design is unfeasible; on the other hand, when the number of the planet gears engaging in series is odd (n is odd), the design is feasible.

9.1.2.2 Two Planet Gears Are Adjacent to Each Other by a Pin-in-Slot Joint

Similarly, the possibility of this condition is checked by inspecting the output of the epicyclic gear train and the hypothetical condition of the epicyclic gear train. The derivation process and its conclusions are presented in Chap. 7. Therefore, when the number of the planet gears engaging in series is even (n is even), the condition is feasible; on the other hand, when the number of the planet gears engaging in series is odd (n is odd), the condition is unfeasible.

All possible assembly situations of the planet gears are discussed, for the epicyclic gear train with more than one planet gear. Based on the discussion, it is understood that the incident links of the pin-in-slot joint closely affect the numbers of members and joints. In addition, the pin-in-slot joint is related to the assembly situations of the planet gears. This is useful for the archaeological work. Once the evidence about the pin-in-slot joint is decoded in the future, the real mechanisms could be confirmed accordingly.

In summary, two design types are concluded based on the concept of simpler structure. Type 1 is a 4-bar mechanism with 5 joints, and Type 2 is a 5-bar mechanism with 7 joints. Furthermore, it is confirmed that the two planet gears of Type 2 design must be adjacent to each other by a pin-in-slot joint.

9.2 Design Process of the Planetary Subsystem

Once again, the reconstruction work applies the design procedure presented in Chap. 5. All feasible reconstruction designs of the two design types of the planetary subsystem are generated. Focusing on the inferior planets or the superior planets, the appropriate driving power arrangements corresponding to the feasible reconstruction designs are identified. Thus, the relations about the tooth calculation are concluded through the kinematic analysis of the reconstruction designs.

9.2.1 Type 1 Design of the Planetary Subsystem

Type 1 design of the planetary subsystem is developed from the planetary theory of Apollonius' epicyclic system. Its output motion has to be the anomaly motion, for either inferior planets or superior planets. The design by Edmunds and Morgan belongs to Type 1 reconstruction design [11]. The purposes and roles of the links are confirmed based on the research of the historical archives. However, due to the designs of epicyclic gear trains that the output rotation direction could not be identified based on the input rotation direction and the number of gear joints, the input link is uncertain when the output rotation is doubtful. Thus, the driving power arrangement is also unknown. Therefore, the design specifications of Type 1 reconstruction design are concluded as follows:

1. It is an epicyclic gear train.
2. It is a 4-bar mechanism with 5 joints and 1 degree of freedom.
3. The joints include 3 revolute joints, 1 external gear joint, and 1 pin-in-slot joint.
4. The types of mechanical members are gears and links.

In the same way as the design process of the solar subsystem, the generalization process is carried out based on the concept of the generalization and the number synthesis of generalized chains. The design by Edmund and Morgan is generalized into the corresponding generalized chain, as shown in Fig. 9.1. In addition, the number of the generalized chains with 4 links and 5 joints is 1, as shown in Fig. 8.3.

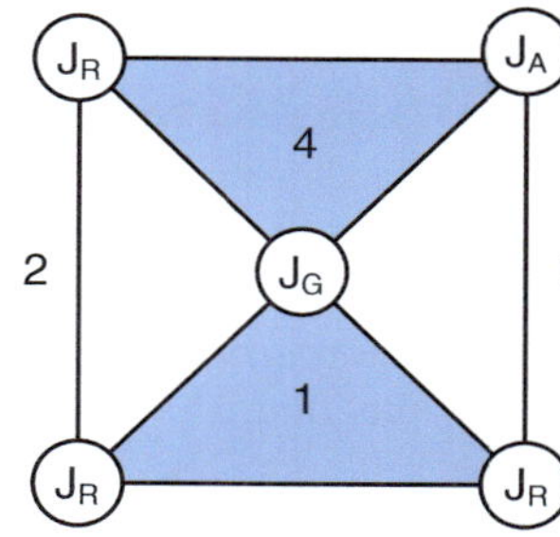

Fig. 9.1 Generalized chain corresponding to the design by Edmund and Morgan

According to the research of the historical archives, the design constraints of the Type 1 reconstruction design for the lost planetary motions are concluded as follows:

1. Based on the kinematic analysis of Apollonius' epicyclic theory for the planetary motions, the mechanism of the planetary subsystem must have the function of rate variation.
2. In order to match the front zodiac dial decoded by Freeth et al., the output link must rotate around axis *b* that represents the motion of date.
3. Based on the previous investigation, the mechanism for a specific planet's motion could be an epicyclic gear train. It must have at least a ground link (member 1, K_F), an input link (member 2, K_I), an output link (member 3, K_O), and a transmission link (member 4, K_{T1}).

   Ground link (member 1, K_F)

   (a) There must be a ground link as the frame. As far as the epicyclic train is considered, it includes at least a sun gear, a carrier, a planet gear, and an output link. And, the output, the planet gear, and the carrier must have revolute joints. Since there is no remaining revolute joint adjacent to the sun gear, the sun gear must be fixed as the ground link.
   (b) Except for the transmission link, each remaining link must be adjacent to the ground link; therefore, the ground link must be a multiple link.

   Input link (member 2, K_I)

   (a) There must be a link as the input link that functions as the carrier.
   (b) The input link must be adjacent to the ground link with a revolute joint.
   (c) The input link must not be adjacent to the output link to avoid the degeneration of the mechanism.

   Output link (member 3, K_O)

   (a) There must be a binary link as the output link.
   (b) The output link must generate the periodic anomaly motion that satisfies the Apollonius' planetary theory.
   (c) The output link must be adjacent to the ground link with a revolute joint.

Transmission link (member 4, K_T)

(a) There must be at least a link as the transmission link that functions as the planet gear.
(b) According to the consideration of the function, the transmission link must be a multiple link with at least 1 revolute joint and 1 gear joint.
(c) The transmission link must not be adjacent to the ground link with a revolute joint.

Pin-in-slot joint (joint e, J_A)

(a) There must have a pin-in-slot joint, functioning as a cam joint, to transmit the motion and generate the necessary function of the variable rate.
(b) The two links incident to the pin-in-slot joint must not rotate around the common axis.
(c) Based on the previous mechanism analysis, the pin-in-slot joint must be incident to the planet gear; therefore, the pin-in-slot joint must be incident to the transmission link.
(d) The pin-in-slot joint must not be incident to the ground link and the input link.

Revolute joint (joints a, b, c; J_R)

(a) There must be 3 revolute joints.
(b) Every link must have at least 1 revolute joint.
(c) Any joint incident to the input link must be a revolute joint.
(d) There can be no loop formed exclusively by revolute joints.

Gear joint (joint d, J_G)

(a) According to the previous mechanism analysis, the ground link of Type 1 design should function as the sun gear of the epicyclic gear train. There must be a gear joint incident to the ground link.
(b) There can be no 3-bar loop formed exclusively by gear joints.

Next, all feasible specialized chains subject to above-concluded design constraints are identified through the following steps:

Step 1 For each available generalized chain, assign the ground link.
Step 2 For each result from Step 1, assign the output link.
Step 3 For each result from Step 2, assign the input link.
Step 4 For each result from Step 3, assign the remaining link as the transmission link.
Step 5 For each result from Step 4, assign the pin-in-slot joint.
Step 6 For each result from Step 5, assign the gear joints.
Step 7 For each result from Step 6, assign the remaining joints as the revolute joints.

For the only possible generalized chain with 4 links and 5 joints shown in Fig. 8.3, its corresponding specialization process is as follows:

1. The ground link is a ternary link. For the generalized chain shown in Fig. 8.3, the two multiple links are symmetric. Therefore, based on the concept of similar classes, the ground link is identified as one of these two multiple links, as shown in Fig. 9.2a.
2. The output link must be a binary link adjacent to the ground link. For the specialized chain with identified ground link shown in Fig. 9.2a, the two binary links adjacent to the ground link is symmetric. Therefore, based on the concept of similar classes, the output link is identified as one of these two binary links, as shown in Fig. 9.2b.
3. Since the input link must be a binary link adjacent to the ground link, there is only one link that can be identified as the input link, as shown in Fig. 9.2c.
4. The transmission link must be a ternary link. Therefore, the remaining link is identified as the transmission link, as shown in Fig. 9.2d.
5. The pin-in-slot joint must be incident to the output link, but not the ground link. Therefore, there is only one result to identify the pin-in-slot joint, as shown in Fig. 9.2e.
6. There is only one gear joint in each design. Since the ground link must be a fixed gear, the gear joint must be incident to the ground link. Furthermore, the input link and the output link must be adjacent to the ground link with revolute joints. Therefore, there is only one joint that can be identified as the gear joint, as shown in Fig. 9.2f.
7. The remaining joints are identified as revolute joints, as shown in Fig. 9.2f.

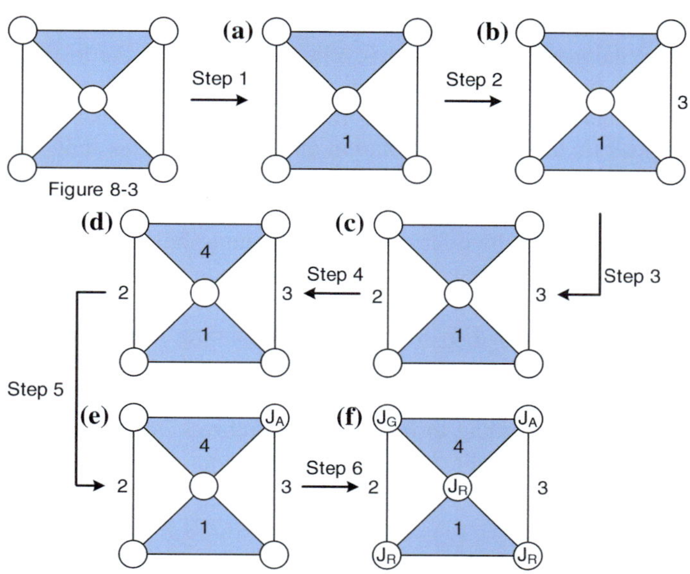

Fig. 9.2 Specialization process of Type 1 design of the planetary subsystem

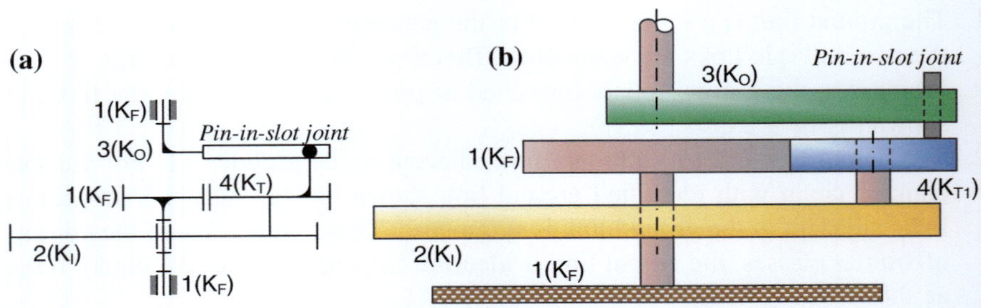

Fig. 9.3 Specialized chain of Type 1 design of the planetary subsystem

Only one specialized chain with the ground link, the output link, the input link, the transmission link, the pin-in-slot joint, 2 revolute joints, and 1 gear joint is available, as shown in Fig. 9.2f.

The specialized chain shown in Fig. 9.2f can be particularized into the corresponding schematic and mechanical drawings, as shown in Fig. 9.3. Based on the concluded kinematic analysis of epicyclic gear train, the feasible driving power arrangements of the design shown in Fig. 9.3 can be identified to generate the reconstruction designs of the planetary subsystem with clockwise rotation on the front dial. Then, for the obtained reconstruction designs, their relations of tooth calculating and suitable planets are determined. However, whatever planets, it is always observed that the output shaft of reconstruction designs to demonstrate the planet pointers must be coaxial with the shafts of the date subsystem and the lunar subsystem.

First, the input link of the design shown in Fig. 9.3 must rotate clockwise, based on the kinematic analysis of epicyclic gear train. By the inspection of the possible driving power arrangements as shown in Fig. 8.1, two of the driving power arrangements can be applied for this design so as to make the reconstruction design generate the clockwise rotation on the front dial, as shown in Fig. 9.4. In the following, all reconstruction designs shown in Fig. 9.4 are presented in detail.

For the simulations of the inferior planets, the input link, as the deferent of the epicyclic system, should rotate at the rate of the tropical year (ω_{year}). Therefore, the input link has to be connected to axis *b* by observing the possible driving power conditions shown in Fig. 8.1. The corresponding reconstruction design is illustrated in Fig. 9.4a. Then, the planet gear incident to the pin-in-slot joint, as the epicycle, should rotate at the rate of the epicycle period (ω_{e}). It is known that the sun gear is fixed according to the historical archives. In the same way for the kinematic analysis of epicyclic gear trains of the solar subsystem, the following expressions deriving the relations of rates above are true.

Fixed sun gear:

$$x + y = 0 \tag{9.1}$$

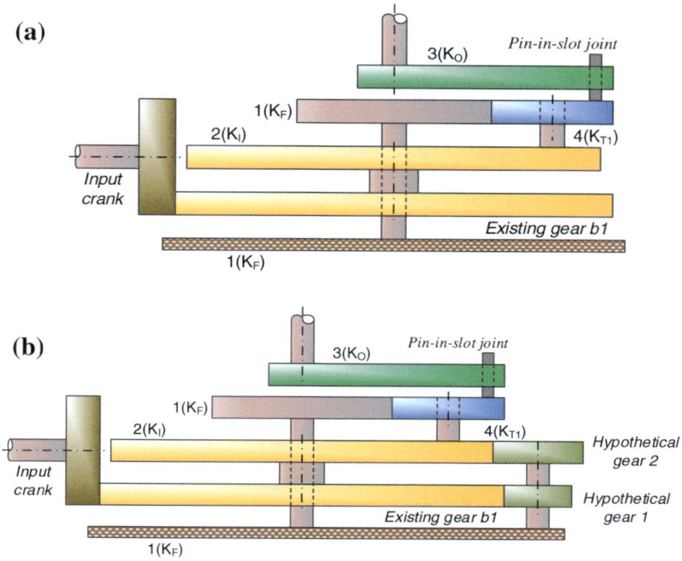

Fig. 9.4 Type 1 feasible designs of the planetary subsystem

Carrier as the deferent:

$$x = \omega_{\mathrm{d}} = \omega_{\mathrm{year}} \tag{9.2}$$

Planet gear as the epicycle:

$$-\frac{T_{\mathrm{s}}}{T_{pn}} y = \omega_{\mathrm{e}} \tag{9.3}$$

By solving Eqs. (9.1)–(9.3), the relation of tooth calculation to simulate the designated inferior planet is concluded as:

$$\frac{T_{\mathrm{s}}}{T_{pn}} = \frac{\omega_{\mathrm{e}}}{\omega_{\mathrm{d}}} = \frac{\omega_{\mathrm{e}}}{\omega_{\mathrm{year}}} \quad \text{(for the inferior planets)} \tag{9.4}$$

In fact, to compare with the reconstruction design shown in Fig. 9.4a, Eq. (9.4) is represented by the numbers of teeth of link 1 and link 4 as:

$$\frac{T_1}{T_4} = \frac{\omega_{\mathrm{e}}}{\omega_{\mathrm{d}}} = \frac{\omega_{\mathrm{e}}}{\omega_{\mathrm{year}}} \quad \text{(for the inferior planets)} \tag{9.5}$$

For the superior planets, the input, as the deferent, should rotate at the rate of the deferent period (ω_{d}). The driving power arrangements could make the rate of the input link changeable. Thus, the corresponding reconstruction design is illustrated in Fig. 9.4b. Furthermore, the planet gear incident to the pin-in-slot joint, as the epicycle, should rotate at the rate of the tropical year (ω_{year}). With the above requirements, the following expressions are true:

Planet gear as the epicycle:

$$x = \omega_{\mathrm{d}} \tag{9.6}$$

Carrier as the deferent:

$$-\frac{T_{\mathrm{s}}}{T_{\mathrm{p}n}} y = \omega_{\mathrm{e}} = \omega_{\mathrm{year}} \tag{9.7}$$

By solving Eqs. (9.1), (9.6), and (9.7), the relation of tooth calculation to simulate the designated superior planet is concluded as:

$$\frac{T_{\mathrm{s}}}{T_{\mathrm{p}n}} = \frac{\omega_{\mathrm{e}}}{\omega_{\mathrm{d}}} = \frac{\omega_{\mathrm{year}}}{\omega_{\mathrm{d}}} \quad \text{(for the superior planets)} \tag{9.8}$$

To compare with the design shown in Fig. 9.4b, Eq. (9.8) is represented by the numbers of teeth of link 1 and link 4 as:

$$\frac{T_1}{T_4} = \frac{\omega_{\mathrm{e}}}{\omega_{\mathrm{d}}} \quad \text{(for the superior planets)} \tag{9.9}$$

Moreover, the relation of tooth calculation to generate the appropriate input rates for the designated superior planet is concluded as:

$$\frac{\omega_2}{\omega_{b1}} = \frac{\omega_{\mathrm{d}}}{\omega_{\mathrm{year}}} = \frac{T_{b1} \times T_{\mathrm{s}2}}{T_{\mathrm{s}1} \times T_2} \tag{9.10}$$

In summary, the one synthesized design of Type 1 design of planetary subsystem shown in Fig. 9.3 can, respectively, generate one feasible result corresponding to the inferior and superior planets by applying its appropriate driving power arrangement. The relations of tooth calculation corresponding to each result are concluded. Figure 9.5 shows these feasible designs in 3D models.

9.2.2 Type 2 Design of the Planetary Subsystem

Type 2 design is doubtful for its feasible driving power arrangements and the possible directions of the input rotations. Based on historical archives, the design specifications of Type 2 design are concluded as follows:

1. It is an epicyclic gear train.
2. It is a 5-bar mechanism with 7 joints and 1 degree of freedom.
3. The joints include 4 revolute joints, 2 external gear joints, and 1 pin-in-slot joint.
4. The types of mechanical members are gears and links.
   The atlas of the generalized chains with 5 links and 7 joints is listed, as shown in Fig. 8.7. The historical archives indicate that Type 2 reconstruction design must have at least a ground link (member 1, K_{F}), an input link (member 2, K_{I}), an output link (member 3, K_{O}), and 2 transmission links (members 4 and 5, K_{T1} and K_{T2}).

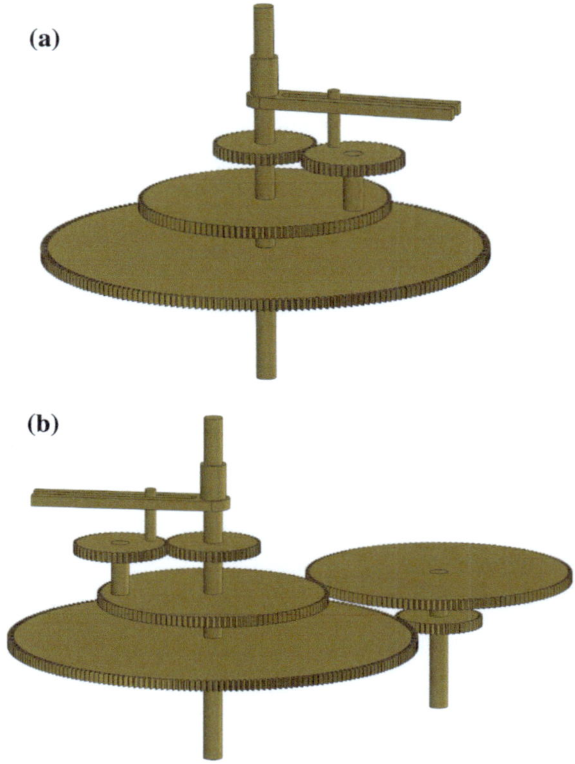

Fig. 9.5 3D models of Type 1 feasible designs of the planetary subsystem. **a** Corresponding to Fig. 9.4a; **b** corresponding to Fig. 9.4b

Ground link (member 1, K_F)

(a) There must be a ground link as the frame.
(b) The ground link must be adjacent to the input link, the output link, and the transmission link as the carrier; therefore, the ground link must be a multiple link.

Input link (member 2, K_I)

(a) There must be a link as the input link.
(b) The input link must be adjacent to the ground link with a revolute joint.
(c) The input link must not be adjacent to the output link to avoid the degeneration of the mechanism.

Output link (member 3, K_O)

The corresponding design constraints of output link for Type 2 are the same as Type 1 mentioned previously.

Transmission links (member 4, K_{T1}; member 5, K_{T2})

(a) There must be two multiple links with at least 1 revolute joint and 1 gear joint, as the transmission links.
(b) There must be at least one transmission link as the planet gear; therefore, it must not be adjacent to the ground link with a revolute joint.
(c) When two transmission links function as the planet gears, they must be adjacent to each other with a pin-in-slot joint.

Pin-in-slot joint (joint g, J_A)

(a) There must have a pin-in-slot joint, functioning as a cam joint, to transmit the motion and generate the necessary function of the variable rate.
(b) The two links incident to the pin-in-slot joint must not rotate around the common axis.
(c) Based on the previous mechanism analysis, the pin-in-slot joint must be incident to the planet gear; therefore, the pin-in-slot joint must be incident to the transmission link.
(d) The pin-in-slot joint must not be incident to the ground link and the input link.

Revolute joint (joints a, b, c, d; J_R)

(a) There must have 4 revolute joints.
(b) Every link must have at least 1 revolute joint.
(c) There can be no loop formed exclusively by revolute joints.

Gear joint (joints e, f; J_G)

(a) There must have 2 gear joints.
(b) There can be no 3-bar loop formed exclusively by the gear joints.
(c) There must be a gear joint incident to the ground link.

The identification steps of links and joints are the same as the steps mentioned in the previous example. Following the steps and the concluded design constraints, all possible specialized chains can be identified. For the three generalized chains with 5 links and 7 joints shown in Fig. 8.7, the specialization process is presented as follows.

9.2.2.1 Ground Link (Member 1)

1. For the generalized chain shown in Fig. 8.7a, the assignment of the ground link generates 2 results, as shown in Fig. 9.6a, b.
2. For the generalized chain shown in Fig. 8.7b, the assignment of the ground link generates 2 results, as shown in Fig. 9.6c, d.
3. For the generalized chain shown in Fig. 8.7c, the assignment of the ground link generates 1 result, as shown in Fig. 9.6e Therefore, 5 specialized chains with identified ground link are available, as shown in Fig. 9.6a–e.

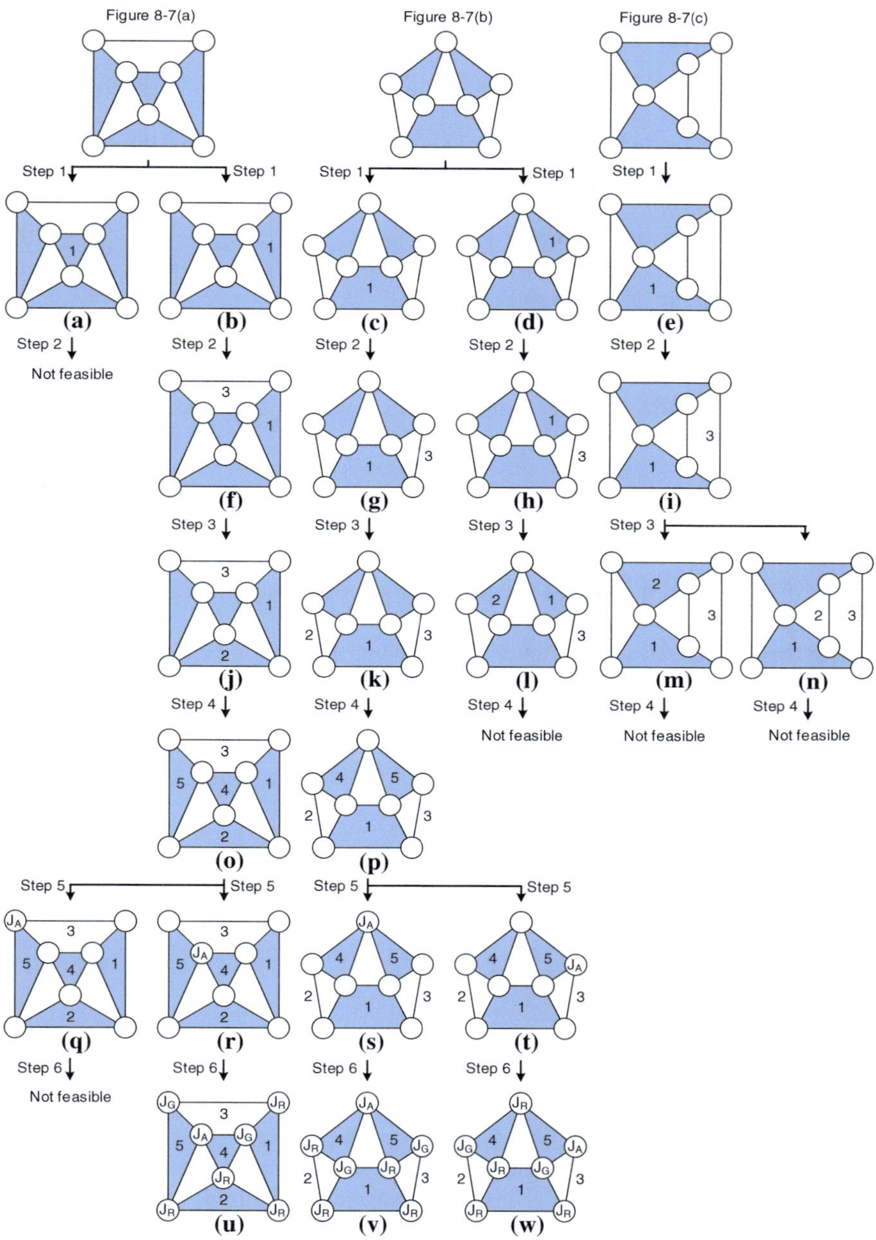

Fig. 9.6 Specialization process of Type 2 design of the planetary subsystem

9.2.2.2 Output Link (Member 3)

1. For the specialized chain shown in Fig. 9.6a, the assignment of the output link generates zero results.
2. For the specialized chain shown in Fig. 9.6b, the assignment of the output link generates 1 result, as shown in Fig. 9.6f.
3. For the specialized chain shown in Fig. 9.6c, the assignment of the output link generates 1 result, as shown in Fig. 9.6g.
4. For the specialized chain shown in Fig. 9.6d, the assignment of the output link generates 1 result, as shown in Fig. 9.6h.
5. For the specialized chain shown in Fig. 9.6e, the assignment of the output link generates 1 result, as shown in Fig. 9.6i.

Therefore, 4 specialized chains with identified ground link and output link are available, as shown in Fig. 9.6f–i.

9.2.2.3 Input Link (Member 2)

1. For the specialized chain shown in Fig. 9.6f, the assignment of the input link generates 1 result, as shown in Fig. 9.6j.
2. For the specialized chain shown in Fig. 9.6g, the assignment of the input link generates 1 result, as shown in Fig. 9.6k.
3. For the specialized chain shown in Fig. 9.6h, the assignment of the input link generates 1 result, as shown in Fig. 9.6l.
4. For the specialized chain shown in Fig. 9.6i, the assignment of the input link generates 2 results, as shown in Fig. 9.6m, n.

Therefore, 5 specialized chains with identified ground link, output link, and input link are available, as shown in Fig. 9.6j–n.

9.2.2.4 Transmission Links (Members 4 and 5)

Since the transmission links must be multiple links, the remaining two links should be multiple links. By the inspection of the 5 specialized chains with identified ground link, output link, and input link shown in Fig. 9.6j–n, only 2 specialized chains shown in Fig. 9.6j, k are qualified to have the two transmission links. Therefore, there are 2 specialized chains with identified ground link, output link, input link, and transmission links, as shown in Fig. 9.6o–p.

9.2.2.5 Pin-in-Slot Joint (Joint J_A)

Once the numbers of the planet gears of the epicyclic gear train are even, the pin-in-slot joint must be incident to the planet gears. In addition, the pin-in-slot joint must not be incident to the ground link and the input link. Thus, the pin-in-slot joint can be identified as follows:

1. For the specialized chain shown in Fig. 9.6o, the assignment of the pin-in-slot joint generates 2 results, as shown in Fig. 9.6q, r.
2. For the specialized chain shown in Fig. 9.6p, the assignment of the pin-in-slot joint generates 2 results, as shown in Fig. 9.6s, t.

Therefore, there are 4 specialized chains with identified ground link, output link, input link, transmission links, and pin-in-slot joint, as shown in Fig. 9.6s, t.

9.2.2.6 Gear Joints (Joint J_G)

There must be two gear joints in each design. One gear joint is incident to the ground link, since the ground link is regarded as a fixed sun gear. The other gear joint can be identified based on the assignment results of the pin-in-slot joint. Once the pin-in-slot joint is incident to the transmission links, the remaining gear joint must be incident to the output link. Thus, based on the constraints, these 2 gear joints can be identified as follows:

1. For the specialized chain shown in Fig. 9.6q, the assignment of the gear joints generates zero results.
2. For the specialized chain shown in Fig. 9.6r, the assignment of the gear joints generates 1 result, as shown in Fig. 9.6u.
3. For the specialized chain shown in Fig. 9.6s, the assignment of the gear joints generates 1 result, as shown in Fig. 9.6v.
4. For the specialized chain shown in Fig. 9.6t, the assignment of the gear joints generates 1 result, as shown in Fig. 9.6w.

   Therefore, 3 specialized chains with identified ground link, output link, input link, transmission links, pin-in-slot joint, and gear joints are available, as shown in Fig. 9.6u–w.

9.2.2.7 Revolute Joints (Joint J_R)

There are 4 revolute joints in each design. Therefore, the remaining joints of these 3 specialized chains shown in Fig. 9.6u–w are identified as revolute joints.

Through the above specialization process, 3 specialized chains with identified ground link, input link, output link, transmission links, pin-in-slot joint, revolute joints, and gear joints are available, as shown in Fig. 9.6u–w. In the following particularization process, the identifications of the proper driving power arrangements and the relations of tooth calculation for each feasible reconstruction design are analyzed:

1. According to the specialized chain shown in Fig. 9.6u, the pin-in-slot joint is incident to two planet gears. For the design of an epicyclic gear train, the

relative kinematic analysis is introduced in Chap. 7. In order to satisfy the required deferent's rate, the rate of the input link should be changeable. Therefore, the specialized chain is particularized into the kinematic drawing by collating its possible driving power arrangement, as shown in Fig. 9.7a. Furthermore, Fig. 9.8a shows this feasible reconstruction design in 3D solid models.

Based on the kinematic analysis of Apollonius' epicyclic system introduced in Chap. 1, the kinematic function of the celestial bodies' revolution on the system is concluded as expressed as Eq. (7.4). It is feasible either the inferior planets or the superior planets. Therefore, by comparing with the analysis result of gear trains introduced in Chap. 7, the following expressions are true:

$$\omega_o = \left\{x + (-1)^n\left(\frac{T_s T_{pk}}{T_{p(k-1)}T_o}\right)y\right\} + F\left\{(-1)^{n+1}\left(\frac{T_s T_{pk}}{T_{p(k-1)}T_o}\right)y\right\} = \omega_{deferent} + F\omega_{epicycle}$$

$$\Rightarrow \begin{cases} x + (-1)^n \dfrac{T_s T_{pk}}{T_{p(k-1)}T_o} y = \omega_d \\ (-1)^{n+1} \dfrac{T_s T_{pk}}{T_{p(k-1)}T_o} y = \omega_e \end{cases} \quad (9.11)$$

For the inferior planets, the rate of the deferent (ω_d) is the rate of the tropical year (ω_{year}). And, the rate of the epicycle is different with the inferior planets. Certainly, the sun gear is fixed. Furthermore, by solving Eqs. (9.1), (9.11), the relation of tooth calculation for the designated inferior planet is concluded as:

$$\frac{(-1)^n T_{p(k-1)}T_o - T_s T_{pk}}{T_s T_{pk}} = \frac{\omega_{year}}{\omega_e} \quad \text{(for the inferior planets)} \quad (9.12)$$

As to the superior planets, the rate of the epicycle (ω_e) is the rate of the tropical year (ω_{year}). The rate of the epicycle is different to the superior planets. Therefore, the relation of tooth calculation is concluded as:

$$\frac{(-1)^n T_{p(k-1)}T_o - T_s T_{pk}}{T_s T_{pk}} = \frac{\omega_d}{\omega_{year}} \quad \text{(for the superior planets)} \quad (9.13)$$

In this design, the number of planet gears is 2 ($n = 2$). Thus, by the inspection of the design shown in Fig. 9.7a, Eqs. (9.12) and (9.13) are represented as:

$$\frac{T_3 T_4 - T_1 T_5}{T_1 T_5} = \frac{\omega_{year}}{\omega_e} \quad \text{(for the inferior planets)} \quad (9.14)$$

$$\frac{T_3 T_4 - T_1 T_5}{T_1 T_5} = \frac{\omega_d}{\omega_{year}} \quad \text{(for the superior planets)} \quad (9.15)$$

Based on Eqs. (9.11) and (9.12), the rate of the input link is concluded as:

$$\omega_2 = \omega_\mathrm{d} + \omega_\mathrm{e} \tag{9.16}$$

Based on the calculation of the gear ratios, the relation between the input link and the surviving gear $b1$ is derived as:

$$\frac{\omega_2}{\omega_{b1}} = \frac{T_{b1} \times T_{s2}}{T_{s1} \times T_2}$$

or

$$\omega_2 = \frac{T_{b1} \times T_{s2}}{T_{s1} \times T_2} \omega_{b1} \tag{9.17}$$

2. Since the output gear meshes the transmission gear incident to the pin-in-slot joint to transmit the anomaly motion, the input rotation must be counterclockwise, i.e., the input link must engage with contrate gear $a1$. By choosing appropriate driving power arrangement, the specialized chain shown in Fig. 9.6v is particularized into the kinematic drawing shown in Fig. 9.7b. Figure 9.8b shows this design in 3D solid model. Finally, by the kinematic analysis of gear trains and the concluded Eq. (7.4), the following expressions about the tooth calculation are concluded:

$$\omega_3 = \frac{T_5}{T_3} \omega_5 \tag{9.18}$$

$$\omega_5 = \omega_\mathrm{d} + F\omega_\mathrm{e} = \omega_2 + F\omega_4 \tag{9.19}$$

$$\frac{T_1}{T_4} = \frac{T_\mathrm{e}}{T_\mathrm{d}} \tag{9.20}$$

$$\frac{\omega_2}{\omega_{b1}} = \frac{T_{b1}}{T_2} \tag{9.21}$$

$$\frac{\omega_\mathrm{e}}{\omega_{b1}} = \frac{T_1 \times T_{b1}}{T_2 \times T_4} \tag{9.22}$$

Although the designs could be applied on both the inferior planets and the superior planets, the real application of the planetary subsystem of Antikythera device should be considered further by the geometric constraints, i.e., space of the interior structure.

3. Since the input link meshes the carrier which determines the direction of the revolution, the input link must rotate counterclockwise. By choosing the proper driving power arrangement, the specialized chain shown in Fig. 9.6w is particularized into the kinematic drawing shown in Fig. 9.7c. Figure 9.8c shows this

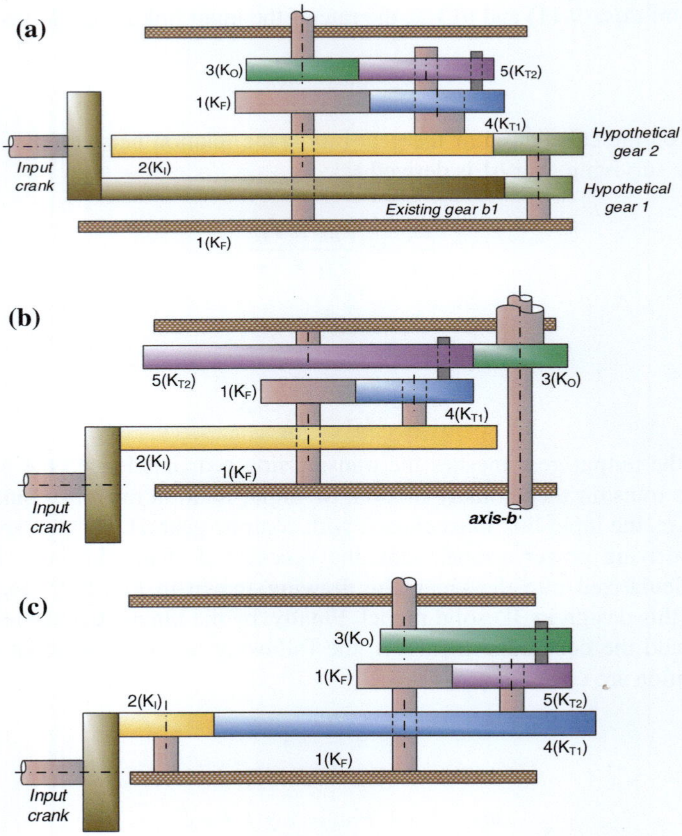

Fig. 9.7 Type 2 feasible designs of the planetary subsystem

design in a 3D solid model. In the same way, the relations of tooth calculation are concluded by the calculation of the gear ratios as:

$$\frac{T_1}{T_5} = \frac{\omega_{\mathrm{e}}}{\omega_{\mathrm{d}}} \tag{9.23}$$

$$\frac{T_4}{T_{b1}} = \frac{\omega_{\mathrm{year}}}{\omega_{\mathrm{d}}}. \tag{9.24}$$

9.3 Remarks

The lost planetary subsystem is reconstructed through the systematic design methodology presented in Chap. 5, and 5 feasible reconstruction designs are generated. The mechanism types of these feasible designs are concluded based on the

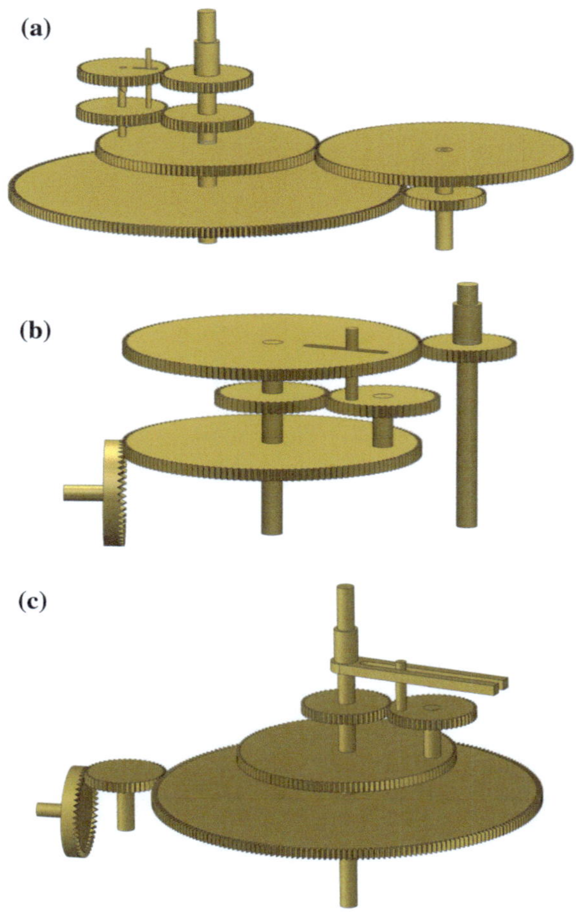

Fig. 9.8 3D models of Type 2 feasible designs of the planetary subsystem. **a** Corresponding to Fig. 9.7a; **b** corresponding to Fig. 9.7b; **c** corresponding to Fig. 9.7c

kinematic and mechanism analyses and known concepts from the study of other astronomical mechanical instruments and ancient astronomy. Nevertheless, as mentioned in Chap. 8 regarding the solar subsystem, all design types are based on the concept that the mechanism is the simpler the better. Similarly, the design types can be changed with more excavated evidence in the future.

Furthermore, it is emphasized that the compound gears are not applied, in order to generate the simpler structure of mechanisms. Such an assumption affects the tooth calculation of the reconstructed planetary subsystem. Indeed, the obtained results can satisfy the periods of planets and generate the appropriate teeth based on the concluded teeth relations. The relevant discussions are introduced in Chap. 11. It has to be understood that the topological structure of mechanisms of each feasible reconstruction design does not change, even if the compound gears are considered in the planetary subsystem. Nevertheless, the design with compound gears really provides more possible conditions of tooth calculation owing to more gear joints (gear ratios). Furthermore, the advantage resulting from the application of compound gears is to generate more accurate output motions to simulate the astronomical theory or the actual astronomical phenomena.

References

1. Wright MT (2002) A planetarium display for the Antikythera Mechanism. Horol J 144(5):169–173
2. Freeth T, Jones A (2012) The Cosmos in the Antikythera Mechanism. In: Institute for the Study of the Ancient World, New York University. Available in the website: Institute for the Study of the Ancient World (ISAW). http://dlib.nyu.edu/awdl/isaw/isaw-papers/4/. Accessed July 2014
3. Evans J, Carman CC, Thorndike AS (2010) Solar anomaly and planetary displays in the Antikythera Mechanism. J Hist Astron 42:1–39
4. Carman CC, Thorndike AS, Evans J (2012) On the pin-and-slot device of the Antikythera Mechanism, with a new application to the superior planets. J Hist Astron 43(150):93–116
5. Evans J, Carman CC (2014) Mechanical astronomy: a route to the ancient discovery of epicycles and eccentrics. In: Sidoli N, Van Brummelen G (eds) Alexandria, through baghdad: surveys and studies in the ancient greek and medieval islamic mathematical sciences in honor of J.L. Berggren. Springer, Berlin, pp 145–171
6. Lin JL, Yan HS (2014) Historical development of reconstruction designs of Antikythera Mechanism. In: Applied mechanics and materials, vol 163, pp 1–6, Mar 2012. Presented in proceedings of the 9th international conference on history of mechanical technology and mechanical design, Tainan, 23–25 Mar 2014
7. Yan HS (2007) Reconstruction designs of lost ancient chinese machinery. Springer, The Netherlands
8. Yan HS (1992) A methodology for creative mechanism design. Mech Mach Theory 27(3):235–242
9. Yan HS (1998) Creative design of mechanical devices. Springer, Singapore
10. Yan HS, Lin JL (2012) Reconstruction synthesis of the lost subsystem for the planetary motions of Antikythera Mechanism. ASME Trans J Mech Design 134:011003. Paper also presented in the 35th mechanisms and robotics conferences, Washington, D.C., USA, 28–31 Aug 2011 (paper no. DETC2011-47773)
11. Edmunds M, Morgan P (2000) The Antikythera Mechanism: still a mystery of greek astronomy. Astron Geophys 41: 6.10–6.17

Chapter 10
Reconstruction Designs of the Moon Phase Display Device

There is a black-and-white ball over the front dial of the Antikythera device. This ball seems to be inlaid into a bronze disk. The device, including the disk and the ball, has been regarded as the moon phase display device. The detailed design regarding how to operate this device by the output motions of the interior mechanism is a quest, due to the damage surviving evidence. The purpose of this chapter is to analyze all feasible designs of the moon phase display device.

Historical developments of the pointer designs of mechanical devices are introduced. All pointers of the Antikythera device are analyzed and classified based on their motions. Furthermore, all feasible designs regarding the unclear pointer design of the moon phase display device are presented by the kinematic and mechanism analyses.

10.1 Historical Archives of the Moon Phase Display Device

In the Antikythera device, the moon phase display device over the front dial is to demonstrate the cyclic variation of the moon phase. The indicating information is communicated through the pointer design in which a black-and-white ball rotates in a hole of disk. In order to exactly restore the moon phase display device, the building of its historical archives aims for the research of the surviving evidence and the feasible driving power arrangements, the analysis of pointers of the Antikythera device, and the mechanism and kinematic analyses, so as to define the possible design types.

J.-L. Lin and H.-S. Yan, *Decoding the Mechanisms of Antikythera Astronomical Device*, DOI 10.1007/978-3-662-48447-0_10

10.1.1 Related Evidence and Available Designs

Figure 4.8a shows the surviving fragment corresponding to the moon phase display device. The visible bronze disk has been regarded as a part of the device. It is evidenced that a partial wreck of contrate gear is hidden in this fragment. Owing to the founding of the contrate gear, it is reasonably to suppose that there should be a gear engaging the contrate gear so as to make the contrate gear meaningful. Therefore, based on the existing research and the related evidence, the moon phase display device is defined to have at least 3 members: a visible bronze disk, a contrate gear connected to the black-and-white ball pointer, and a hypothetical gear engaging the contrate gear. In addition, the device has been confirmed as an orthogonal gear train.

Wright and Freeth et al. restored the moon phase display device in their reconstruction designs [1–4]. Furthermore, Freeth et al. presented two different designs. Whatever the available reconstruction designs of this device, all of them are epicyclic gear trains with two degrees of freedom. The main concept of these available designs is that the change rate of the moon phase is based on the rate difference between the daily motion (solar motion) and the lunar motion.

Wright's reconstruction design is the same as the first design of Freeth et al. in 2006, as shown in Fig. 10.1 [1, 4]. The epicyclic gear train consists of 4 links, 3 revolute joints, and 1 gear joint. The sun gear is connected to the output shaft indicating the date. The bronze disk, which is visible in the surviving evidence, is connected to the output shaft demonstrating the lunar motion, and it is the carrier of the epicyclic gear train. Therefore, based on the kinematic analysis of epicyclic gear trains, the contrate gear, i.e., the planet gear of the epicyclic gear train, can rotate once in a month around its axis. Thus, for the whole moon phase display device, its hand has a revolution with the rate of a month around the center of the front dial and a rotation with the rate of a month.

In 2008, another reconstruction design by Freeth et al. revealed a different epicyclic gear train to demonstrate the cyclic variation of moon phase, as shown in Fig. 10.2 [2, 5]. The two designs by Freeth et al. have opposite orientations on the contrate gears of the moon phase display device. Owing to this difference, the

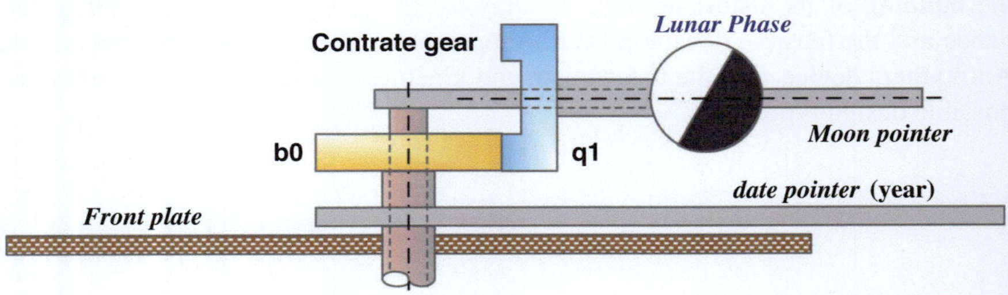

Fig. 10.1 Reconstruction design of the moon phase display device by Wright [4] and Freeth et al. [1]

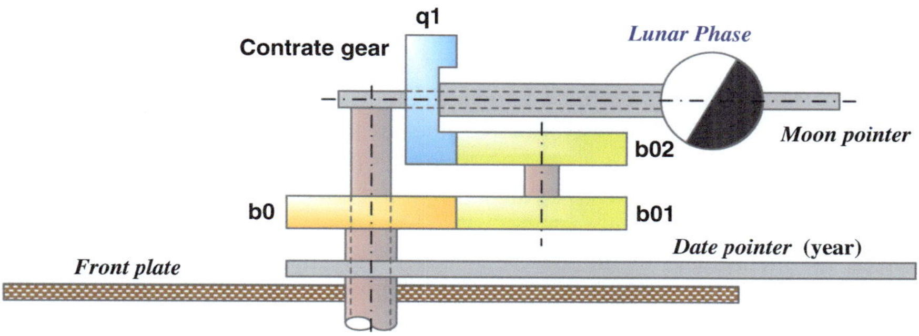

Fig. 10.2 Reconstruction design of the moon phase display device by Freeth et al. [2, 5]

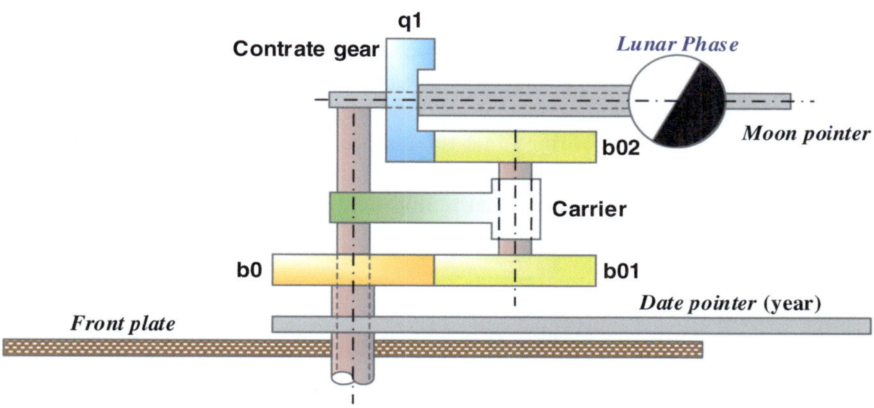

Fig. 10.3 Improved reconstruction design of the moon phase display device by Efstathiou et al. [6]

second design has a planet gear functioning as an idle gear, except for the surviving contrate gear. This idle planet gear is hypothetical. Nevertheless, the surviving bronze disk does not obviously appear in this reconstruction design, even though it is understood that this disk is the carrier. In 2012, Efstathiou et al. [6] improved the unclear illustration to show the carrier for this epicyclic gear train of the moon phase display device, and they built the operation model, as shown in Fig. 10.3.

The second reconstruction design by Freeth et al. is an epicyclic gear train including 5 links, 4 revolute joints, and 2 gear joints. According to the kinematic analysis of epicyclic gear trains, regarding the two designs by Freeth et al., it is found that the rotation directions of the contrate gears on the bronze disk (carrier) are contrary. Under the condition of the same rotation rates, the hand's rotation direction of the second design is opposite to the hand's rotation direction of the first design.

10.1.2 Possible Driving Power Arrangements

The moon phase display device is known as the most upper device, and it is powered by the output coaxial shafts of the interior mechanism. Its available reconstruction designs mentioned above are logical. All of them are with 2 degrees of freedom. In fact, the related surviving evidence is unclear and cannot provide sufficient information to distinguish the assembly situations. The topological structure of the mechanism of this device is uncertain. Therefore, there can be other different designs that satisfy the confirmed information from the surviving evidence. Here, based on the experiences of other reconstruction works of the interior subsystems, it is reasonable to suggest that the reconstruction designs of the interior subsystems can be mechanisms with 1 or 2 degrees of freedom. Furthermore, the possible driving power arrangements are concluded by direct inspection. For these two designs, the required driving power sources have to be the motions of the innermost one shaft or the innermost two of the coaxial shafts.

However, the designs regarding how to assemble the lost planetary subsystem and the lost solar subsystem are uncertain. For the sequence of the coaxial pointers over the front dial, it is confirmed that the innermost pointer is the Moon pointer. However, the pointer next to the Moon pointer is uncertain. Such an uncertain condition closely depends on the reconstruction designs of the moon phase display device.

10.1.3 Possible Design Types

The possible numbers of links and joints is a question to reconstruct the moon phase display device. The bronze disk is a key factor to resolve this question. According to the surviving evidence, the function of the bronze disk cannot be confirmed, even though the existing reconstruction designs logically indicate that the disk is the carrier. The designs that the bronze disk is rotatable or fixed must be identified further. Its function can result in different mechanism types of the moon phase display device. Accordingly, the following two cases are presented:

1. When the bronze disk is fixed, the moon phase display device is an ordinary gear train.
2. When the bronze disk is rotatable, the moon phase display device is an epicyclic gear train.

An ordinary gear train must be a mechanism with 1 degree of freedom. The derivation process of relations about the numbers of links and joints is the same as that of the calendrical subsystem presented in Chap. 6. Based on Eqs. (6.4)–(6.6), the numbers of links and joints can be calculated. For the designs of ordinary gear trains, the simplest structure is a 3-bar mechanism with 2 revolute joints and 1 gear joint.

Furthermore, the designs of epicyclic gear trains can be mechanisms with 1 or 2 degrees of freedom. For an epicyclic gear train with N_L members, N_{JR} revolute joints, and N_{JG} gear joints, based on mobility analysis, the following expressions are true, respectively, for mechanisms with 1 degree of freedom and 2 degrees of freedom:

$$\text{1-DOF: } 2N_{JR} + N_{JG} - 3N_L + 4 = 0 \tag{10.1}$$

$$\text{2-DOF: } 2N_{JR} + N_{JG} - 3N_L + 5 = 0 \tag{10.2}$$

By solving Eqs. (10.1), (6.2) and (6.3) in terms of the number of links, the relations about the number of joints for 1-DOF epicyclic gear trains are the same as Eqs. (6.4)–(6.6). Nevertheless, by solving Eqs. (10.2), (6.2), and (6.3) in terms of the number of links, the following relations about the numbers of joints for 2-DOF epicyclic gear trains are concluded:

$$N_J = 2N_L - 4 \tag{10.3}$$

$$N_{JR} = N_L - 1 \tag{10.4}$$

$$N_{JG} = N_L - 3 \tag{10.5}$$

Therefore, if the moon phase display device is an epicyclic gear train with 1-DOF, its simplest structure is a 3-bar mechanism with 2 revolute joints and 1 gear joint. If this device is an epicyclic gear train with 2-DOF, the simplest structure is a 4-bar mechanism with 3 revolute joints and 1 gear joint.

Three design examples are presented. Design Example 1 is an ordinary gear train with 3 links, 2 revolute joints, 1 gear joint, and 1-DOF. Design Example 2 is an epicyclic gear train with 3 links, 2 revolute joints, 1 gear joint, and 1-DOF. Design Example 3 is an epicyclic gear train with 4 links, 3 revolute joints, 1 gear joint, and 2-DOF.

10.2 Design Process of the Moon Phase Display Device

In this section, all possible designs of the three concluded design examples of the moon phase display device are synthesized. However, the reconstruction work of the moon phase display device is slightly different from the ones of the interior subsystems presented in Chaps. 6–9. The structures of all concluded design examples are simple. Whatever 3-bar or 4-bar mechanisms, the number of the generalized chain for each mechanism is 1, as shown in Fig. 10.4. Therefore, the feasible specialized chain can be obtained by direct inspection.

In what follows, the three design examples of the moon phase display device are illustrated. The whole design of mechanisms of the coaxial pointers over the front dial can be known clearly based on the obtained designs of the device. How to assemble the moon phase display device with the rotation shafts of the coaxial pointers can be clearly understood. The purpose of the bronze disk can be

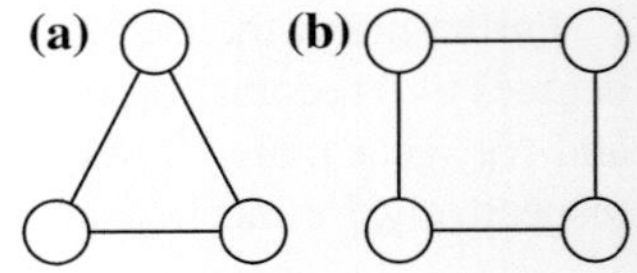

Fig. 10.4 Atlases of **a** (3, 3) and **b** (4, 4) generalized chains

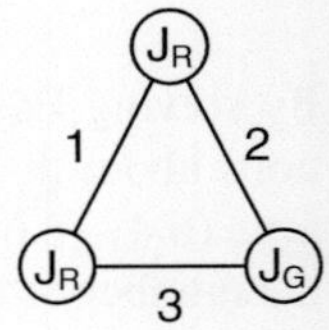

Fig. 10.5 A specialized chain of the moon phase display device—Example 1

explained. Meanwhile, it is resolved that the black-and-white ball pointers are with revolution or without revolution, except for the rotation.

10.2.1 Example 1: Ordinary Gear Trains

Firstly, based on the historical archives, the design specifications of Example 1 are described as follows:

1. It is a 3-bar mechanism with 2 joints and 1 degree of freedom.
2. The mechanical members are gears, including a contrate gear.
3. The joints consist of 2 revolute joints and 1 external gear joint.
4. The bronze disk has to be fixed as the frame.
5. The gear train of the moon phase display device is an ordinary gear train, i.e., all gears rotate around the fixed axes.

The mechanism includes the ground link (link 1, K_F), an input link (link 2, K_I), and an output link (link 3, K_O). The only 1 generalized chain shown in Fig. 10.4a is specialized into the corresponding specialized chain, as shown in Fig. 10.5. The bronze disk has to be fixed as the frame in this example. Therefore, based on the known requirements of members and the concluded design specifications, the mechanical drawing of the possible design of Example 1 is generated, as shown in Fig. 10.6. The input link is the sun gear and connected by the shaft of the Moon pointer, and it rotates once in a month (ω_{month}). The output link is the contrate gear, i.e., the black-and-white ball pointer.

The design shown in Fig. 10.6 is different to the previous research [1–4, 6]. Since the bronze disk is fixed, it cannot be the Moon pointer. It is understood that there should be 8 pointers (for the Sun, the Moon, Mercury, Venus, Mars, Jupiter, Saturn, and the date of Egyptian calendar) with axial rotation over the front dial.

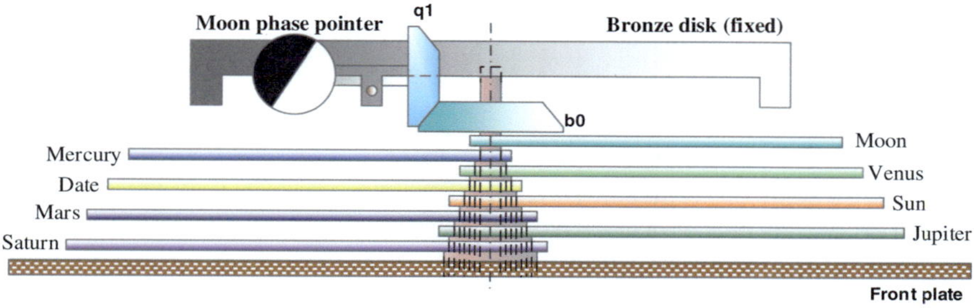

Fig. 10.6 A feasible design of the moon phase display device—Example 1

Also, for the moon phase display device, its black-and-white ball pointer is with a radial rotation but without an axial rotation.

For the design shown in Fig. 10.6, based on its kinematic analysis and the known input arrangement, the following expression is true:

$$\omega_3 = \frac{T_2}{T_3}\omega_{\text{month}} \tag{10.6}$$

Since the contrate gear has to rotate once in a month (ω_{month}), based on Eq. (10.6), it is known that the teeth number of existing gear $a1$ has to be equal to the teeth number of the contrate gear. Furthermore, a bronze top with cup shape (or convex shape) appeared in most existing designs is mechanically connected to the bronze disk.

In summary, the pointer types next to the Moon pointer and the assembly ways of the interior lost subsystems are not considered. This design is suitable for all reconstruction designs.

10.2.2 Example 2: Epicyclic Gear Trains with 1-DOF

The design specifications of Example 2 are described as follows:

1. It is a 3-bar mechanism with 2 joints and 1 degree of freedom.
2. The mechanical members are gears, including a contrate gear.
3. The joints consist of 2 revolute joints and 1 external gear joint.
4. The bronze disk functions as a carrier.
5. The gear train of the moon phase display device is an epicyclic gear train.

Members of the mechanism include the ground link (link 1, K_F), an input link (link 2, K_I), and an output link (link 3, K_O). The only generalized chain shown in Fig. 10.4a is specialized into the corresponding specialized chain, as shown in Fig. 10.7. Since the bronze disk is rotatable as the carrier, only the input link (link 2) can be assigned to the bronze disk. In addition, the output link (link 3) is always the contrate gear connected to a black-and-white ball pointer. Thus, in order to satisfy the requirements of the known mechanical members and the design

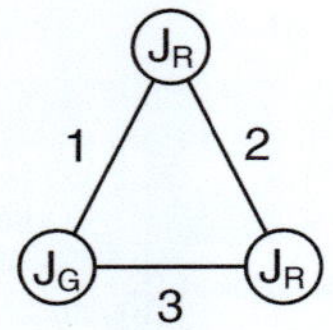

Fig. 10.7 A specialized chain of the moon phase display device—Example 2

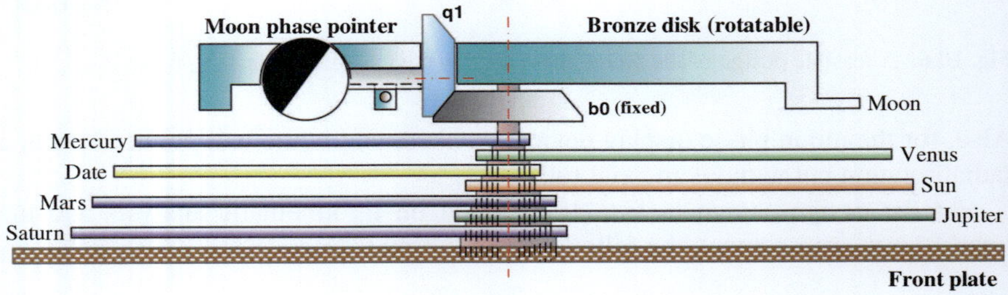

Fig. 10.8 A feasible design of the moon phase display device—Example 2

Table 10.1 Kinematic analysis of the design shown in Fig. 10.8

Motion conditions	Gear types		
	Sun gear ω_s	Carrier ω_c	Planet gear ω_p
All members rotate at x_1 rpm	x_1	x_1	x_1
Carrier is fixed, and sun gear rotates at y_1 rpm	y_1	0	$\left(\frac{T_s}{T_p}\right)y_1$
Absolute rotational speed	$x_1 + y_1$	x_1	$x_1 + \left(\frac{T_s}{T_p}\right)y_1$

specifications, the obtained specialized chain shown in Fig. 10.7 is particularized into its mechanical drawing, as shown in Fig. 10.8. The input link, the bronze disk, is connected to the output shaft of the Moon pointer, i.e., the disk rotates once in a month. The existing gear $a1$, which is the sun gear of the epicyclic gear train, has to be fixed as a part of the frame.

The kinematic analysis of an epicyclic gear train including a sun gear, a planet gear, and a carrier is listed in Table 10.1. For the design shown in Fig. 10.8, it is confirmed that the contrate gear must rotate once in a month (ω_{month}), the input link rotates once in a month (ω_{month}), and the sun gear must be fixed. Therefore, the following expressions are true:

$$\frac{T_s}{T_p}y_1 = \omega_{\text{month}} \tag{10.7}$$

$$x_1 + y_1 = 0 \tag{10.8}$$

$$x_1 = \omega_{\text{month}} \tag{10.9}$$

By solving Eqs. (10.7)–(10.9), it is confirmed that the teeth of existing gear *a*1 must be equal to the teeth of the contrate gear. The bronze disk is mechanically connected to the bronze top with a cup shape to cover the contrate gear. The pointer of the moon phase display device rotates around not only the radical direction but also the axial direction, like the revolution and the rotation in astronomy.

In summary, the shaft next to the shaft of the Moon pointer has to be fixed. For this reason, the application of the reconstruction design of Example 2 should be considered by the reconstruction design of the interior structure of mechanisms.

10.2.3 Example 3: Epicyclic Gear Trains with 2-DOF

The mean variation rate and the true variation rate are two important concepts to show the variation cycle of the moon phase. Firstly, the moon phase finishes its cyclic variation during a month in average. This average variation rate almost approximates the rate of the lunar motion. Therefore, the concept to utilize the rate of the lunar motion to express the average variation rate of the moon phase is feasible and has been applied in Examples 1 and 2. However, it is known that such a concept is not accurate to demonstrate the moon phase well and truly.

In fact, based on the geocentric model, the moon phase results from the relative positions between the Earth, the Moon, and the Sun. The cycle of the moon phase is related to the cycles of the solar motion and the lunar motion. It is known that the cyclic variation rate of the moon phase can be derived from the difference between the cyclic rates of the Moon and the Sun. Thus, the variation rate of the moon phase is regarded as the true variation rate of the moon phase. This concept about the true variation rate of the moon phase is applied in this example. Therefore, in order to demonstrate the true variation rate of the moon phase, the rates of the solar motion and the lunar motion are necessary for the reconstruction design. It is logically that the pointer next to the Moon pointer should be the Sun pointer. Thus, two required input rates of this design are confirmed. One input rate is the cyclic rate of the lunar motion. The other input rate is the rate of the solar motion. These two input sources are the sun gear and the carrier of the epicyclic gear train, respectively.

In the same reasoning as Examples 1 and 2, the related design specifications are defined in accordance with the historical archives as follows:

1. It is a 3-bar mechanism with 2 joints and 2 degrees of freedom.
2. The mechanical members are gears, including a contrate gear.
3. The joints consist of 2 revolute joints and 1 external gear joint.
4. The bronze disk functions as a carrier.
5. The gear train of the moon phase display device is an epicyclic gear train.

The mechanism contains the ground link (link 1, K_F), 2 input links (links 2 and 3, K_{I1} and K_{I2}), and an output link (link 4, K_O). The feasible generalized chain is shown in Fig. 10.4b. Based on direct inspection, 2 specialized chains are obtained,

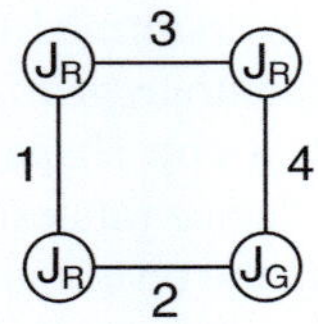

Fig. 10.9 A specialized chain of the moon phase display device—Example 3

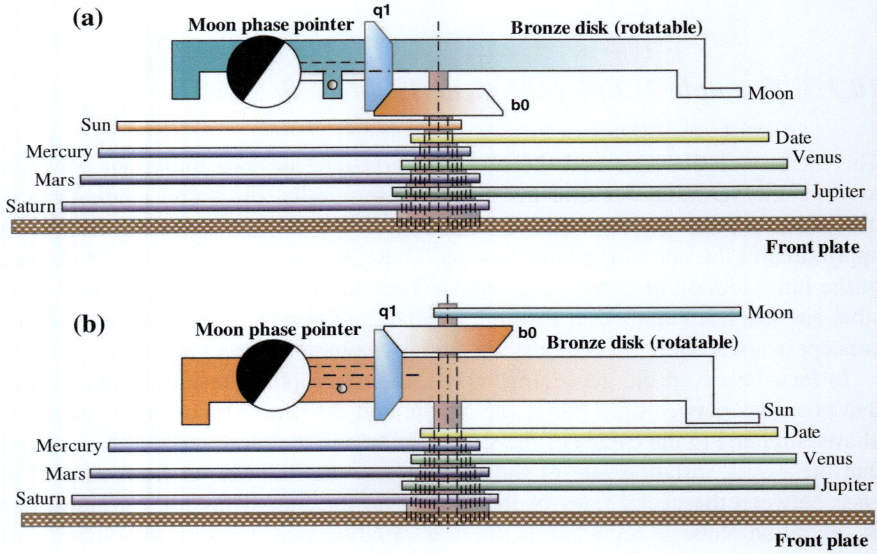

Fig. 10.10 Two feasible designs of the moon phase display device—Example 3, **a** carrier is the Moon pointer, **b** carrier is the Sun pointer

as shown in Fig. 10.9. Then, by the particularization process, 2 corresponding mechanical drawings can be generated, as shown in Fig. 10.10. Whatever designs, it is consistent that the innermost of the coaxial shaft rotates once in a month to demonstrate the lunar motion.

In the design shown in Fig. 10.10a, the bronze disk connects to the innermost of the coaxial shafts. Thus, its sun gear, i.e., the other input source, must connect to the shaft of the Sun pointer. Based on kinematic analysis listed in Table 10.1, the following expressions are true:

$$\omega_{\mathrm{s}} = x_1 + y_1 = \omega_{\mathrm{sun}} \tag{10.10}$$

$$\omega_{\mathrm{c}} = x_1 = \omega_{\mathrm{moon}} \tag{10.11}$$

$$\frac{T_{\mathrm{s}}}{T_{\mathrm{p}}} y_2 = \frac{T_{\mathrm{s}}}{T_{\mathrm{p}}} (\omega_{\mathrm{sun}} - \omega_{\mathrm{moon}}) \tag{10.12}$$

However, in the designs shown in Fig. 10.10b, the innermost of the coaxial shafts connects to the hypothetical sun gear. Furthermore, its carrier must connect to the shaft of the Sun pointer. Based on kinematic analysis listed in Table 10.1, the following expressions are true:

$$\omega_{\mathrm{s}} = x_1 + y_1 = \omega_{\mathrm{moon}} \tag{10.13}$$

$$\omega_{\mathrm{c}} = x_1 = \omega_{\mathrm{sun}} \tag{10.14}$$

$$\frac{T_{\mathrm{s}}}{T_{\mathrm{p}}} y_1 = \frac{T_{\mathrm{s}}}{T_{\mathrm{p}}} (\omega_{\mathrm{moon}} - \omega_{\mathrm{sun}}) \tag{10.15}$$

Equations (10.12) and (10.15), respectively, introduce the contrate gear to rotate around its shaft in their designs, i.e., the rotation of the moon phase pointer on the bronze disk.

By solving Eqs. (10.10) and (10.12), it is concluded that the teeth of the contrate gear equals to the teeth of the hypothetical sun gear. Also, this result can be derived by solving Eqs. (10.13) and (10.15). The design shown in Fig. 10.10a is the design by Wright and Freeth et al. Both the reconstruction designs shown in Fig. 10.10a, b reveal that the hand of the moon phase device rotates not only around the shaft of the ball pointer but also around the center of the front dial.

Furthermore, the mean rate of the solar motion is once in a year. The cycle of the solar motion can approximate the cycle of annual motion. The requirements of the solar motion can be replaced by the annual motion, i.e., the other input rate is the annual rate. Since the motion of the date pointer based on the Egyptian calendar is the annual motion, the pointer next to the Moon pointer can be the date pointer. Certainly, the replacement based on the concept of approximation results in slight inaccuracy of the moon phase display device.

In a word, 2 feasible designs of Example 3 are obtained: One is the existing design and the other is an alternative design. In accordance with the concept of approximation, the input with the rate of the solar motion can replace the input with the rate of the annual motion. Even though such an approximate condition makes the moon phase display device slightly inaccurate, it is possible for the design of the moon phase display device.

References

1. Freeth T et al (2006) Decoding the ancient Greek astronomical calculator known as the Antikythera mechanism. Nature 444:587–589
2. Freeth T et al (2008) Calendars with olympiad display and eclipse prediction on the Antikythera mechanism. Nature 454:614–661
3. Wright MT (2005) Counting months and years: the upper back dial of the Antikythera mechanism. Bull Sci Instr Soc 87:8–13
4. Wright MT (2006) The Antikythera mechanism and the early history of the moon phase display. Antiquarian Horology 29(3):319–329

5. Freeth et al T (2008) Calendars with olympiad display and eclipse prediction on the Antikythera mechanism. Nature 454:614–617. (Supplementary Notes)
6. Efstathiou K, Basiakoulis A, Efstathiou M, Anastatsiou M, Seiradakis JH (2012) Determination of the gears geometrical parameters necessary for the construction of an operational model of the Antikythera mechanism. Mech Mach Theory 52:219–231

Chapter 11
Assembly Work and Models

Chapters 6–9 introduce the reconstruction work of the incomplete calendrical subsystem, the lunar subsystem with the unexplained feature, the lost solar subsystem, and the lost planetary subsystem, respectively. All feasible reconstruction designs of each interior subsystem are generated by the systematic design procedure presented in Chap. 5, subject to the concluded design constraints and the standard of technique in the time of the Antikythera device.

This chapter introduces the assembly work of the interior mechanisms in accordance with the assembly procedure of 3D reconstruction models. Forty-eight feasible reconstruction designs of the complete interior structure of the device composed of the six subsystems are generated through the identification of geometrical constraints, design of mechanical members, and evaluation of space arrangement. Furthermore, the appropriate designs of moon phase display device corresponding to these reconstruction designs with the complete interior mechanisms are provided. Thus, all mechanisms composed of the Antikythera astronomical device are decoded completely. Finally, by assembling the interior mechanisms, the moon phase display device, and the exterior appearance, a 3D reconstruction model of the Antikythera device is built.

11.1 Complete Interior Mechanisms

The reconstruction works presented in Chaps. 6–9 show all feasible reconstruction designs of the unclear and lost subsystems. In addition, several reconstruction designs combining with various driving power arrangements are obtained for the solar subsystem and the planetary subsystem. For the calendrical subsystem, two feasible designs are generated. For the lunar subsystem, two feasible designs are generated. For the solar subsystem, four designs are generated with three design types. For the planetary subsystem, four designs are obtained with two design

J.-L. Lin and H.-S. Yan, *Decoding the Mechanisms of Antikythera Astronomical Device*, DOI 10.1007/978-3-662-48447-0_11

types, and all of them are feasible for the five planets. The complete interior mechanism is composed of the surviving structure of the confirmed subsystem and the reconstruction designs of the unclear and lost subsystems. Thus, there are a total of 16,384 feasible designs of the complete interior mechanism without considering the space arrangement of the subsystems and the constraints of the gear sizes. This large number of designs is almost helpless for restoring the lost ancient mechanism. Consequently, in order to converge the amounts of all feasible designs with the complete interior mechanisms and build appropriate reconstruction models, some constraints for the assembly work of these subsystems are identified.

11.1.1 Assembly Constraints of the Lost Mechanisms

The lost mechanisms are the furriest parts of the interior design. The assembly work of the corresponding reconstruction designs is much flexible. It should be discussed further by considering geometric conditions and mutual effects of the lost subsystems. Here, focusing on the lost mechanisms including the solar subsystem and the planetary subsystem, the assembly constraints contributed to reduce the amount of designs are introduced as follows.

11.1.1.1 Driving Power of Lost Mechanisms

Choosing an appropriate driving power arrangement is an important factor for the reconstruction designs of the lost subsystems. As mentioned in Chaps. 8 and 9, three possible driving power arrangements are identified, as shown in Fig. 8.1. In addition, the resulting reconstruction designs collocate with their appropriate driving power arrangements. Nevertheless, all of these driving power arrangements show that the driving power of the lost subsystems is transmitted from the existing interior mechanism, i.e., the confirmed date subsystem. For the lost subsystems including several gear mechanisms, respectively, for the Sun and the five planets, a mechanism driven by another mechanism is not considered. These gear mechanisms contained in the lost subsystems are parallel.

Once the gear mechanisms in the lost subsystems are combined in series, the assembly work is more complicated. The amount of the feasible reconstruction designs of the lost subsystems is too many to analyze. Therefore, the constraint that the reconstructed gear mechanisms of the lost subsystems are parallel about the driving power of the lost subsystems is logical.

11.1.1.2 Gear Sizes

Gear sizes are closely related to their teeth. For the Antikythera device, the teeth must satisfy the designated gear ratios, and the gear sizes must be constrained by the geometric conditions, such as the known positions and the limited space of the mechanical members. In each reconstruction design of the subsystems, the relations of teeth calculations are concluded by kinematic analysis in Chaps. 6–9. Hence, the geometric constraints are introduced further based on the teeth calculated by the concluded teeth relations.

In the design shown in Fig. 8.5a, the transmission link (link 4) must rotate at the rate of the tropical year in order to demonstrate the solar motion. Since the input link (link 2) is an idle gear, according to the kinematic analysis of gear trains, the teeth of transmission link (link 4) must equal to the teeth of the surviving gear $b1$, i.e., the teeth of transmission link (link 4) is 224. The size of the transmission link (gear) is the same as the size of the surviving gear $b1$. Nevertheless, it is known that the geometrical constraints that the rotational shaft of the output link (link 3) to display the solar motion and the rotational shaft of the surviving gear $b1$ (axis b) are coaxial. Thus, based on the center distance of meshing gears, the size of the transmission link (gear) estimated by its teeth cannot satisfy this geometric constraint, even though the size of the input link (idle gear, link 2) is very small.

In fact, the design shown in Fig. 8.5a can be feasible by using a compound gear. Once the input link is a compound idle gear, the appropriate size (teeth) of the transmission link can be estimated by the kinematic analysis of gear trains so as to satisfy the geometric constraint of coaxial rotational shafts. Certainly, the transmission link with 224 teeth is unnecessary. However, corresponding to this design with a compound gear, its defect is the increasing of the number of mechanical members (gears). In other words, the design in Fig. 8.5a is feasible but waste more materials. Whatever the viewpoints of mechanical designs or manufacturing, this is not a practical way to construct the solar subsystem. Therefore, the design in Fig. 8.5a will not be used to reconstruct the solar subsystem in the assembly work of the complete interior mechanism.

Furthermore, the design shown in Fig. 9.7c cannot serve for the superior planets. The reason is similar to what was mentioned above. In order to demonstrate the motions of the superior planets, the rate of the transmission link (link 4, carrier) should equal to the rate of the deferent. Based on Eq. (9.25), the teeth of the transmission link must be the teeth of the surviving gear $b1$ (224 teeth) times the rate of the deferent. Thus, according to the historical records of planets' cycles listed in Table 11.1, it is known that the teeth of the transmission link is many times of the teeth of gear $b1$ (224 teeth), i.e., the size of the transmission link is very large. As a result, the space of the mechanism is not enough to contain such a big gear.

11.1.1.3 Types of Planets

The previous work about ancient astronomy indicates that the known five planets, Mercury, Venus, Mars, Jupiter and Saturn, are divided into two types: the interior planets and the superior planets. For both types, the planetary motions are described by the same geometric model, as shown in Fig. 1.9. The difference is the cycles (periods) of the deferent and the epicycle.

Table 11.1 Data of the epicyclic system [2, 5–7]

Planets	Data			
	Deferent		Epicycle	
	Period (tropical year)	Radius (arbitrary unit)	Period (tropical year)	Radius (arbitrary unit)
Mercury	1	60	0.241	22.5
Venus	1	60	0.615	43.16
Mars	1.88	60	1	39.5
Jupiter	11.862	60	1	11.5
Saturn	29.458	60	1	6.5

Figure 1.9 is the geometric model of universe based on the geocentric theory. The superior planets including Mars, Jupiter, and Saturn are outside the Sun in the geometrical model of universe, and the inferior planets including Mercury and Venus are inside the Sun. Therefore, this work suggests logically that the same kind of planets should adopt the consistent reconstruction designs, i.e., two inferior planets adopt the same designs, and three superior planets adopt the same designs.

11.1.1.4 Epicyclic System of Superior Planets

Again, the driving power arrangements are discussed. In the reconstruction work of lost subsystems, all possible ways to generate the driving power from the surviving structure (date subsystem) are identified. And, the relations of teeth and the rates of the epicycle and the deferent for each reconstruction design of the lost subsystems are concluded. However, in the analysis process of each design, the rates of the epicycle and the deferent are described in parameters, not in datum. The analysis results are in general for each planet. It is not emphasized what planets the reconstruction designs demonstrate, i.e., the reconstruction designs of the lost subsystems are established when a single planet is considered. Based on the historical records, different planets have different rates of the epicycle and the deferent. Once all planets are considered simultaneously for the assembly work of the complete interior mechanisms, it must be identified whether the synthesized reconstruction designs can be used to demonstrate each planet.

The design shown in Fig. 9.7b cannot serve for the superior planets. As far as the epicyclic system of the superior planets is concerned, their deferent have different rates. In each mechanism for one specific superior planet, its input link must have different rates. As mentioned above, the superior planets should be demonstrated with a common design. Once the design shown in Fig. 9.7b is used to demonstrate the superior planets, its contrate gear $a1$ cannot engage with more than one gear to generate the required rates of the superior planets' deferent.

According to the above analysis for the geometric constraints of the lost subsystems, it is concluded that the numbers of feasible reconstruction designs,

Table 11.2 Numbers for feasible combinations of the lost mechanisms

Planets		Sun		
		Feasible designs of the solar subsystem		
Feasible designs of the inferior planets	Feasible designs of the superior planets	Figure 8.5b	Figure 8.9	Figure 8.14b
Figure 9.4a	Figure 9.4b	No. 01	No. 02	No. 03
Figure 9.4a	Figure 9.7a	No. 04	No. 05	No. 06
Figure 9.7a	Figure 9.4b	No. 07	No. 08	No. 09
Figure 9.7a	Figure 9.7a	No. 10	No. 11	No. 12
Figure 9.7b	Figure 9.4b	No. 13	No. 14	No. 15
Figure 9.7b	Figure 9.7a	No. 16	No. 17	No. 18
Figure 9.7c	Figure 9.4b	No. 19	No. 20	No. 21
Figure 9.7c	Figure 9.7a	No. 22	No. 23	No. 24

respectively, for the solar motion, the inferior planetary motion, and the superior planetary motion are three, four, and two, respectively. In the assembly work of the complete interior mechanisms, the solar subsystem can be built by the designs shown in Figs. 8.5b, 8.9, and 8.14b; and the planetary subsystem for the inferior planets can be built by the designs shown in Figs. 9.4a and 9.7a–c; the planetary subsystem for the superior planets can be built by the designs shown in Figs. 9.4b and 9.7a.

Therefore, there are 24 (=3 × 4 × 2) feasible combinations to construct the lost structures (including the solar subsystem and the planetary subsystem). Table 11.2 lists the feasible designs and the corresponding code names of all feasible combinations in detail.

11.1.2 Assembly Work

Here, for the assembly work of the complete interior mechanisms, the feasible reconstruction designs of the remaining subsystems are reaffirmed definitely. As far as the lunar subsystem is concerned, two feasible designs of the lunar subsystem can provide alternative viewpoints to analyze the unexplained feature. Nevertheless, the design by Freeth et al., shown in Fig. 7.9a is universally accepted in the academic community. Therefore, in the assembly work of the interior mechanisms, the lunar subsystem still applies this most feasible design. As to the calendrical subsystem, both of these two synthesized reconstruction designs shown in Fig. 6.9 are applied for the assembly work of the interior mechanisms.

In summary, there are forty-eight feasible reconstruction designs of the complete interior mechanisms with the functions of the calendrical calculation and the displays of the lunar motion, the solar motion, the interior planetary motion, and the superior planetary motion. The complete interior mechanism of each subsystem and their corresponding mechanical drawings are provided in Appendix A.

This work adopts the reconstruction design model (MODEL 9 in Appendix A) shown in Fig. 3.9 to build the 3D model. Except for the available date subsystem and the eclipse subsystem, this complete design consists of the design shown in Fig. 9.7a for the superior planetary motion, the design shown in Fig. 9.4a for the interior planetary motion, the design shown in Fig. 8.9 for the solar motion, the design shown in Fig. 7.9a for the lunar motion, and the design shown in Fig. 6.9a for the calendrical calculation. Focusing on the lost mechanisms of the reconstruction design model, the mechanisms for their designated celestial bodies' motions do not affect each other. Even one of these mechanisms is removed, the remaining mechanisms still operate smoothly. Such a characteristic makes the reconstruction design model much flexible. This is the reason that the reconstruction design model is chosen to build its 3D model. In other words, if a new evidence indicates that the Antikythera device only demonstrates some of the Sun, the inferior planets, the superior planets in the future, the reconstruction design model is still feasible for an original design by directly taking out the mechanisms for the invalid functions.

11.2 3D Reconstruction Model

It is well known that the Antikythera device is a rectangular solid with an approximate size of 315-190-100 mm^3. Several reconstruction models with different exterior appearances have been presented, as the inscriptions cover the mechanism are gradually decoded. The reconstruction design with the most complete decoded information is built by Freeth et al. In Chap. 3, the exterior appearance based on the research by Freeth et al. is described. For its characteristic dials, their numbers, forms, and functions are also introduced.

Nevertheless, the precise positions of the dials on the front and back plates are not indicated clearly in the decoded information, even though the schemes of the reconstructed dials with a scale are given. Therefore, for the reconstruction work without sufficient information of the exterior appearance, the interior reconstruction mechanisms are helpful to locate these exterior dials on the plates. The centers of these circular dials are aligned with the output shafts of the interior subsystems corresponding to these dials. And, the relative directions of these dials are consistent with the design by Freeth et al.

As far as the reconstruction design with the complete structure of mechanism is concerned, its mechanical members have been confirmed in the previous research [1]. All gears of the date subsystem, the lunar subsystem, and the eclipse prediction subsystem are detected. For the incomplete calendrical subsystem, only two gears are the surviving evidence and the remaining gears are unknown. As to the solar and the planetary subsystems, all members of these two subsystems are lost.

Hence, several works have to be done before building the simulation model of the reconstruction design, even the physical model. First, for each subsystem, the detail designs of mechanical members must be done. The characteristics of meshing gears should be determined based on the concept of modern mechanisms.

Then, the 3D interior structure is built based on the approximately relative directions of the exterior dials. Once all gears of the 3D interior mechanisms are located, the information is feed back to position the exterior dials precisely.

11.2.1 Tooth Calculation

In order to establish the simulation or physical model, a systematic assembly procedure is presented for the complete interior reconstruction design, as shown in Fig. 11.1. Focusing on the reconstruction design model, the detail designs of mechanical members of the calendrical subsystem, the solar subsystem, and the

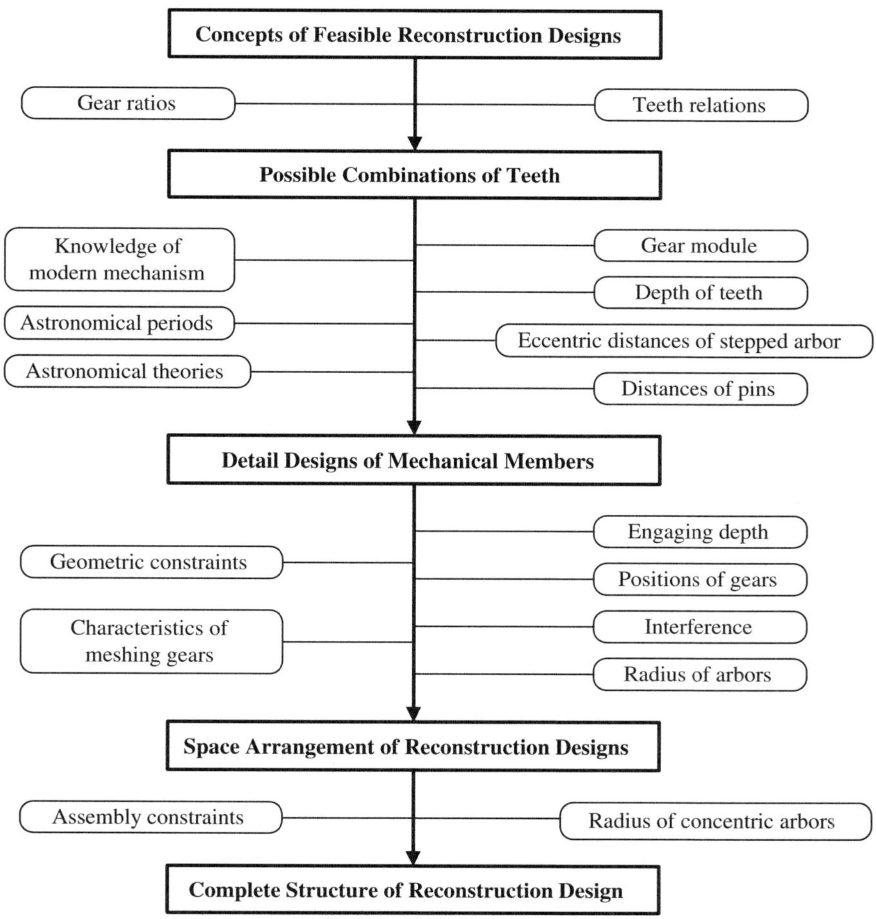

Fig. 11.1 Assembly process of the interior reconstruction design

planetary subsystem are proposed. The teeth calculation work of each subsystem must be done first on the basis of the relations of the concluded teeth relations presented in Chaps. 6–9. By the concept of modern mechanisms, the modulus of gears is defined to deduce the gear sizes further. Meshing gears have the same modulus, and there are several gears detected in the surviving evidence. For the gears estimated by the teeth calculation relations, their modulus is defined arbitrarily by referring to the modulus of the surviving gears. Thus, the sizes of gears can be generated to identify the problems about the interference and arrangements of members in a 3D model.

11.2.1.1 Calendrical Subsystem

The known situations of the calendrical subsystem include the confirmed gear ratios (three display astronomical cycles), two gears, respectively, with 60 teeth and 15 teeth as well as the concluded relations of teeth calculation. These conditions are able to generate reconstruction models. But the amount of the teeth calculation results based on the teeth known situations is numerous for a reconstruction design. Thus, it seems that the problem regarding the accurate teeth calculation of the reconstruction model cannot be adequately resolved.

In order to appropriately estimate all teeth of each feasible design, the following constraints are assumed for sizing the gears by referring to the direct evidences and the design by Freeth et al.:

1. The gear ratio of a gear joint should be 5 at most.
2. The speculative teeth should appear in the existing discovery. The teeth of the reconstruction design are constrained between 10 and 100. Thus, these reconstructed gears are convenient for manufacturing and can fit the space of the mechanism.

Furthermore, according to Eqs. (6.7), (6.8), and (6.9) and the above-mentioned constraints of gear sizes, eight possible results of teeth numbers are generated as follows:

$$(T_4^{46}, T_6^{46}, T_5^{56}, T_6^{56}) = (12, 60, 57, 15)$$

$$(T_4^{46}, T_6^{46}, T_5^{56}, T_6^{56}) = (15, 57, 60, 12)$$

$$(T_4^{46}, T_6^{46}, T_5^{56}, T_6^{56}) = (12, 48, 57, 12)$$

$$(T_4^{46}, T_6^{46}, T_5^{56}, T_6^{56}) = (12, 57, 48, 12)$$

$$(T_4^{46}, T_6^{46}, T_5^{56}, T_6^{56}) = (15, 60, 57, 12)$$

$$(T_4^{46}, T_6^{46}, T_5^{56}, T_6^{56}) = (12, 57, 60, 12)$$

$$(T_4^{46}, T_6^{46}, T_5^{56}, T_6^{56}) = (24, 96, 57, 12)$$

$$(T_4^{46}, T_6^{46}, T_5^{56}, T_6^{56}) = (12, 57, 96, 24)$$

All teeth numbers can be found from the surviving information. The evaluation work of feasible teeth is related to the constraints of gear sizes defined arbitrarily. Hence, once any new evidence is found from the surviving fragments in the future, a group of more reliable teeth can be provided by the same process of teeth calculation.

11.2.1.2 Solar Subsystem

The teeth calculation work of the lost solar subsystem completely depends on the concluded relations of teeth calculation in Chap. 9. For the reconstruction design model, the solar subsystem adopts the design shown in Fig. 8.9. The sufficient space for the reconstruction design must be considered during the process of teeth calculation. Besides, the eccentric distance of the pin-in-slot device is a critical factor to achieve the anomaly function. This eccentric distance is related to the eccentric system of the solar motion. According to the kinematic analysis, the anomaly function by a pin-in-slot device is the anomaly function deduced from the eccentric system. Hipparchus indicated that the eccentric distance is about 1/24 the radius of the circular path, for the eccentric system of the solar motion. Therefore, it is known that the eccentric distance between two links adjacent to each other with a pin-in-slot joint approximates 1/24 the radius of the pin.

Furthermore, in order to generate such an eccentric distance, the center distance between no. 19 gear and no. 20 gear and the center distance between no. 21 gear and no. 22 gear have to be different to result in the eccentric distance. When the modulus of gears is considered, gear sizes corresponding to each group of the feasible teeth are generated. Thus, once the radius of the pin is set in accordance with the size of no. 22 gear (with the pin), the resulting eccentric distance can be evaluated to approximate the ratio of the eccentric distance and the radius of the pin deduced from the ancient astronomical theory, i.e., the ratio is 1/24.

To conclude all above, the teeth calculation work of the lost solar subsystem should satisfy the concluded Eq. (8.25), the space constraint, and the resulting eccentric distance simultaneously. In addition, it takes a lot of work to evaluate a set of feasible teeth. Through try and error and considering the surviving gears like the gears of the lunar subsystem, two sets of teeth and their resulting eccentric distance and radius of pin are as follows:

$$(T_2, T_3, T_4, T_5) = (51, 50, 51, 50)$$

$$(T_2, T_3, T_4, T_5) = (61, 60, 61, 60)$$

Here, the second set of teeth is adopted to build a 3D model, because its result approximates the requirement of the ancient astronomy.

11.2.1.3 Planetary Subsystem

The design of mechanical members of the planetary subsystem should be taken in the same way as the solar subsystem. Teeth calculation, space constraints, and eccentric distance are three important factors. In addition, based on the concluded relations of teeth calculation, the teeth depend on the rates of the epicycle and the deferent. However, the teeth calculation work of the planetary subsystem is not as convenient as the solar subsystem. Even though Apollonius and Hipparchus, who were the astronomers living before or in the era of the Antikythera device and used a famous epicyclic system to describe the planetary anomaly motions, they did not leave definite historical records about the parameters of the geometric model in the historical development of astronomical achievements. For their geometric models, these important parameters, including the radius of the period of the epicycle and the deferent radius, are unclear.

Logically, it is believed that the planets' periods are known in the era and applied for Apollonius' epicyclic system. These periods can be estimated based on the numerical records of astronomical observation, like Babylonian mathematical astronomy. Here, it is especially claimed that the planetary data of Ptolemy's model, as listed in Table 11.1, are used to build the 3D mechanism of the lost subsystem of the reconstruction design model. The related historical records are extracted from "Almagest" [2, 3], the only surviving comprehensive ancient treatise on astronomy.

The Ptolemy's model is slightly different from the Apollonius' model for the planetary motions. The application of the Ptolemy's planetary datum results in the error of teeth calculation. In this model, the resulting teeth calculation is merely an approximate result. Therefore, the reconstruction model emphasizes the configuration design of the lost planetary subsystem and is a conceptual model. Once the required parameters (the radius of the period of the epicycle and the deferent radius) of the epicyclic system for the planetary motions are deduced from the historical literatures, the teeth calculation of the lost subsystem is more appropriate. After the above explanation, the detail designs of all mechanical members are introduced in what follows.

First, for the inferior planets, the teeth calculation is resolved by Eq. (9.5). Two sets of feasible teeth, respectively, for Mercury and Venus are estimated arbitrarily as follows:

$$\text{Mercury} \quad (T_1, T_2, T_4) = (92, 224, 22)$$

$$\text{Venus} \quad (T_1, T_2, T_4) = (64, 224, 40)$$

Then, by Eqs. (9.16) and (9.18) and Ptolemy's planetary datum listed in Table 11.1, three sets of approximate teeth, respectively, for Mars, Jupiter, and Saturn are determined arbitrarily as follows:

$$\text{Mars}\quad (T_1,\, T_2,\, T_3,\, T_4,\, T_5,\, T_{s1},\, T_{s2}) = (36, 215, 54, 96, 94, 20, 30)$$

$$\text{Jupiter}\quad (T_1,\, T_2,\, T_3,\, T_4,\, T_5,\, T_{s1},\, T_{s2}) = (90, 215, 62, 63, 60, 30, 31)$$

$$\text{Saturn}\quad (T_1,\, T_2,\, T_3,\, T_4,\, T_5,\, T_{s1},\, T_{s2}) = (50, 215, 55, 94, 100, 30, 30)$$

All the above teeth are listed in Appendix B. For each planet, these teeth are not unique solutions. Following the above descriptions and the relevant teeth relations, teeth contained in each design can be calculated and adjusted. Again, it has to claim that the detail designs of the lost subsystems are approximate results because of the lost evidence and the application of Ptolemy's planetary datum. Once these two problems are resolved, the origin mechanisms of the lost subsystems can be confirmed.

11.2.2 Detail Designs of Gears

Most of the surviving gears of the Antikythera device are damaged and incomplete. For a complete feasible reconstruction design of the interior mechanism, several gears are lost. Therefore, the restoring work of these incomplete and lost gears must be based on the knowledge of modern mechanisms.

It is undoubted to reconstruct the 3D models of surviving gears based on their measuring data [1]. However, the reconstruction work is hard for the lost gears. Here, the gear module, that is the basis for determining the size of tooth, is used. It is the concept of teeth size and the relation between the diameter of gears and the teeth number. For modern gears, the gear module is the ratio of the diameter of pitch circle and the teeth number [4]. The teeth shape of these reconstructed gears is triangular, unlike involute curve or cycloidal curve of modern gears. Therefore, by approximating the definition of modern gear module, the following equation is proposed for the gears of Antikythera device:

$$D_{\mathrm{a}} = m \times T \tag{11.1}$$

where m is the gear module, T is the number of teeth, and D_{a} is the diameter of the average circle.

And thus, the gear modulus of all detected gears is calculated. For the biggest gear (gear $b1$), its gear module is estimated at about 0.579. In fact, the module of all gears with detection data is estimated [1]. It is found that the modulus of these gears are slightly different, owing to the measuring deviations. This information is used to build the lost gears as references. Once the gear module is determined, the gear size is estimated in accordance with the definition of gear model.

Fig. 11.2 3D reconstruction model of an estimated gear

Furthermore, it is determined that the teeth depth is about 1.2 mm according to the statistical datum of the surviving gears. As mentioned above, the radius of the addendum circle (the diameter of gear to tooth tip) is approximately regarded as the average radius. The radius of the dedendum circle (the radius measured from the center to the tooth bottom) is the average radius minus 1.2 mm. Therefore, the sizes of the essential circles for constructing the gears and their teeth are determined. The solid models of lost gears can be constructed successfully. Figure 11.2 shows the reconstruction model of a lost gear that has a slot in the planetary subsystem and in the solar subsystem. Its teeth number is 60, and the corresponding gear modulus is set at 0.55.

Owing to the definitions of modulus and the teeth depth, the resulting gear sizes are generated, as listed in Appendix B. Now, the center distance between meshing gears is also estimated. In addition, the distance of pin on a gear and the eccentric distance of a stepped arbor are resolved. These problems are important for designing the anomaly motions of heavenly bodies.

The design for the inferior planets is built based on the eccentric system. Taking the mechanism of Mercury for example, the carrier and the planet gear play the deferent and the epicycle, respectively. Thus, by the design shown in Fig. 3.9, the center distance of no. 30 gear (sun gear) and no. 25 gear (planet gear) is the deferent radius, and the radius of the pin over no. 25 gear is the epicycle radius. It is known that the ratio of the epicycle radius and the deferent radius depends on the distance of the pin and the center distance of no. 30 gear and no. 25 gear. Therefore, for the mechanisms of Mercury and Venus, the radii of the pins have to be estimated based on the Ptolemy's planetary datum and the calculated center distance, respectively, as follows:

Mercury:

center distance = 32.775 mm, pin radius = 12.29 mm

Venus:

center distance = 29.9 mm, pin radius = 21.5 mm

In the reconstruction design model, the demonstration of the superior planets applies the design, which is different from the design of the inferior planets. This design shown in Fig. 9.7a is similar to the lunar subsystem. The planet gear with a pin and a slot, respectively, is a significant factor of the design. As mentioned in the kinematic analysis of the lost subsystem and the detail designs of the solar subsystem, the radius of epicycle of the epicyclic system is regarded as the eccentric distance between the planet gear with a pin and the planet gear with a slot. In addition, the deferent radius is regarded as the radius of the pin. Thus, according to the Ptolemy's planetary datum listed in Table 11.1 and gear sizes, the design of members of mechanisms for the superior planets including the radius of pin and the eccentric distance are deduced, respectively, as follows:

Mars:

$$\text{eccentric distance} = 6.25\ \text{mm},\ \text{pin radius} = 9.4937\ \text{mm}$$

Jupiter:

$$\text{eccentric distance} = 1.8125\ \text{mm},\ \text{pin radius} = 9.4565\ \text{mm}$$

Saturn:

$$\text{eccentric distance} = 1.225\ \text{mm},\ \text{pin radius} = 11.308\ \text{mm}$$

In summary, the detail designs of the mechanical members are carried out through the kinematic analysis of each subsystem and the research of the ancient astronomical theories mentioned in Chaps. 1 and 6–9. All teeth, gear sizes, and radius of pin of the reconstruction design model are determined, as listed in Appendix B. Furthermore, the corresponding 3D model is built based on the design information.

11.2.3 Space Arrangement

Space arrangement of the 3D interior mechanisms is based on the positions of gear shafts on the 2D plate that are determined by the characteristic of meshing gears. For a pair of meshing gears, the engaging depth affects the center distance so as to determine the relative positions of the shafts. Therefore, the engaging depth is set at 1.1 mm. There is a clearance about 0.1 mm to achieve the smooth operation of gears. Figure 11.3 illustrates the reconstruction model of two meshing gears.

Based on the reconstruction design model shown in Fig. 3.9, there are six plates to construct the mechanism. Among these plates, the fifth plate is the most complicated one of the 3D model. The positions of gear shafts satisfy the direction of the exterior dials, and their precision affects the smooth operation of gear trains. The orientation work of gear shafts on the remaining plate is relatively easy. Once the orientation work of gear shafts on the fifth plate is completed successfully, the orientation work on the remaining five plates is produced by corresponding to the positions of the gear shafts on the fifth plate.

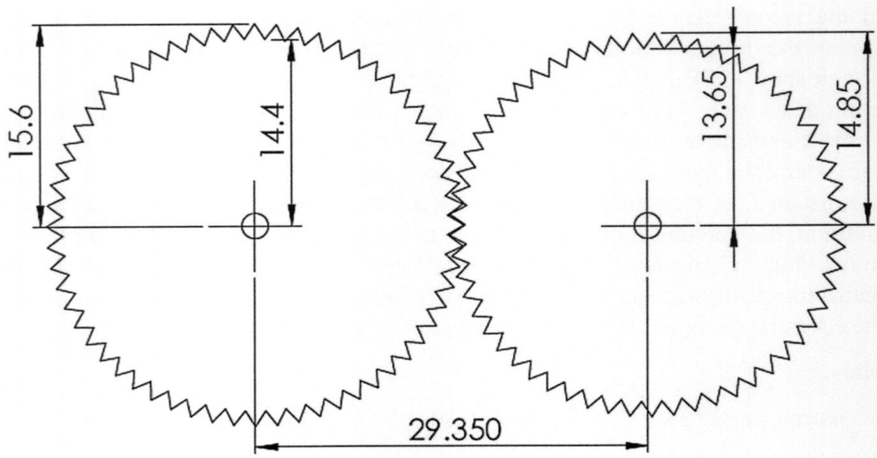

Fig. 11.3 Meshing condition of two reconstructed gears

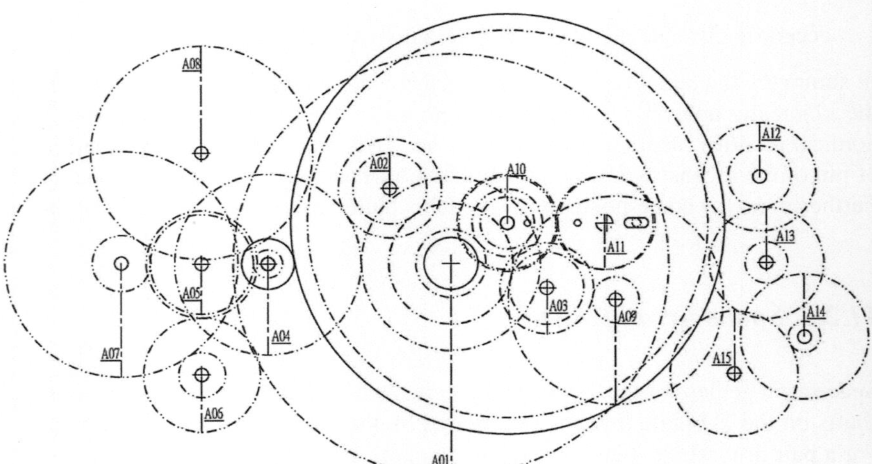

Fig. 11.4 Gear positions of the fifth plate

The relative directions and the approximate positions of the exterior back dials are known. Except for the decoded information of the exterior appearance, the orientation work of the gear shafts of the fifth plate completely depends on the graphic method. After evaluating the states about space arrangement and link interference, the plate that contains the calendrical subsystem, the lunar subsystem, and the eclipse prediction subsystem is generated, as shown in Fig. 11.4.

Fig. 11.5 One 3D model of Yan and Lin's forty-eight reconstruction designs

11.2.4 Simulation Model

Considering the ability of manufacturing and assembly, the positions of gears on their shafts are determined. In addition, the height of the complete mechanism can satisfy the surviving evidence by choosing the appropriate distance above or under gears. Finally, by following the assembly procedure shown in Fig. 11.1 step by step, a 3D model of the reconstruction design model is presented, as shown in Fig. 11.5. In this reconstruction model, the wooden case cover of the Antikythera device is removed in order to show clearly the reconstructed interior gear trains. Certainly, the remaining forty-seven feasible reconstruction designs with complete interior mechanisms can be built based on the same assembly procedure.

Indeed, the building of the 3D simulation model is very helpful in describing the complete concept of Antikythera astronomical calculator. Its animation clearly shows the operation of the mechanism and the demonstration of the pointers on the dials. In a word, the work of building 3D mechanisms is able to precisely prove the feasibility of the obtained reconstruction designs.

References

1. Freeth et al T (2006) Decoding the ancient Greek astronomical calculator known as the Antikythera mechanism. Nature 444:587–891. (Supplementary Notes)
2. Fitzpatrick R (2010) A modern almagest: an updated version of Ptolemy's model of the solar system. The University of Texas (online)
3. Wright MT (2007) The Antikythera mechanism reconsidered. Interdisciplinary Sci Rev 32(1):27–43
4. Yan HS, Wu LI (2014) Mechanisms (in Chinese), 4th edn. Dong Hua Books, Taipei
5. Linton M (2004) From Eudoxus to Einstein: a history of mathematical astronomy. Cambridge University, New York

6. Dreyer JLE (1953) A history of astronomy from Thales to Kepler. Dover Publications Inc., USA
7. Ptolemy C (100) Almagest. English edition: C. Ptolemy (1998) Ptolemy's Almagest (trans: G.J. Toomer). Princeton University Press, USA

Appendix A
All 48 Feasible Designs of Complete Interior Mechanisms

Model 1

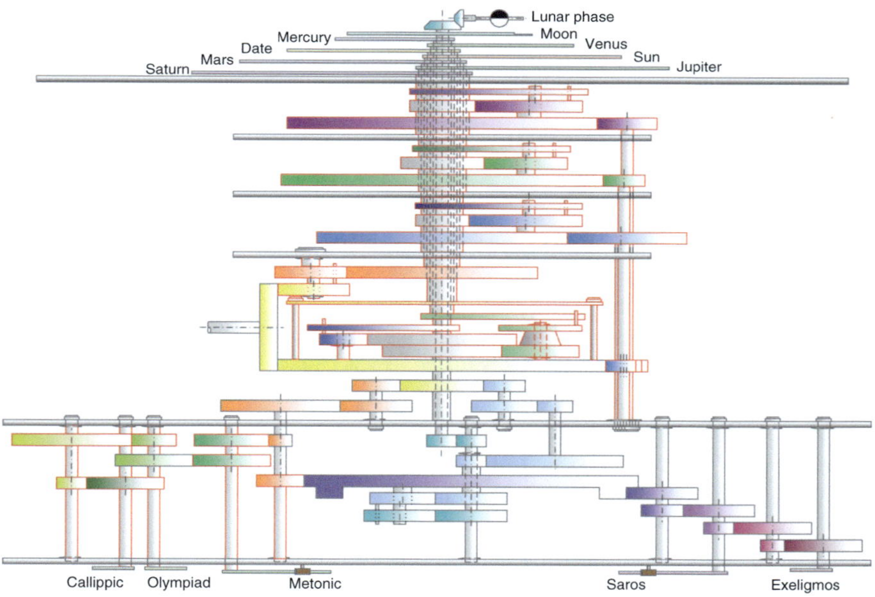

Date subsystem: Design of Fig. 5.7
Eclipse prediction subsystem: Design of Fig. 5.8
Calendrical subsystem: Design of Fig. 6.9a
Lunar subsystem: Design of Fig. 7.9a
Solar subsystem: Design of Fig. 8.5b
(Inferior) Planetary subsystem: Design of Fig. 9.4a
(Superior) Planetary subsystem: Design of Fig. 9.4b
Moon phase display device: Designs of Figs. 10.6 and 10.8

J.-L. Lin and H.-S. Yan, *Decoding the Mechanisms of Antikythera Astronomical Device*, DOI 10.1007/978-3-662-48447-0

Model 2

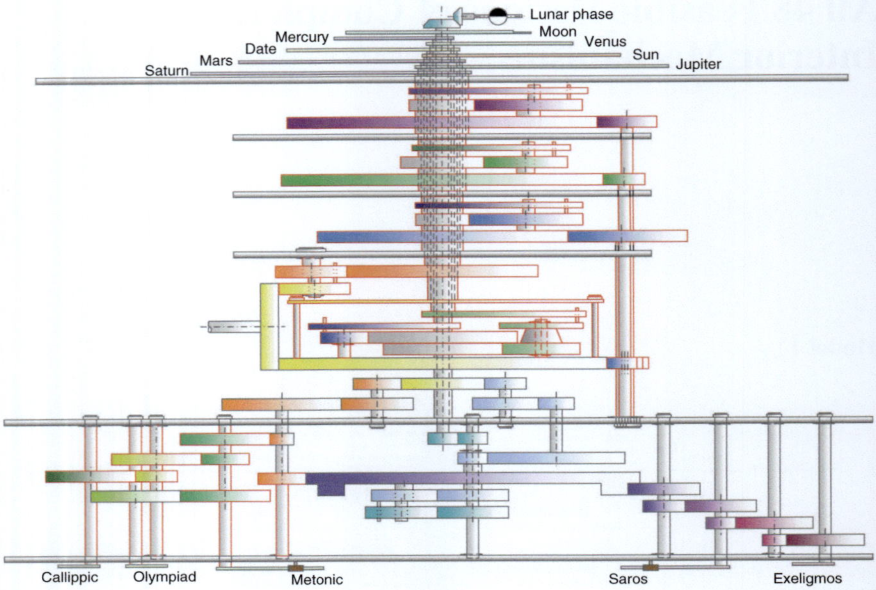

Date subsystem: Design of Fig. 5.7
Eclipse prediction subsystem: Design of Fig. 5.8
Calendrical subsystem: Design of Fig. 6.9b
Lunar subsystem: Design of Fig. 7.9a
Solar subsystem: Design of Fig. 8.5b
(Inferior) Planetary subsystem: Design of Fig. 9.4a
(Superior) Planetary subsystem: Design of Fig. 9.4b
Moon phase display device: Designs of Figs. 10.6 and 10.8

Model 3

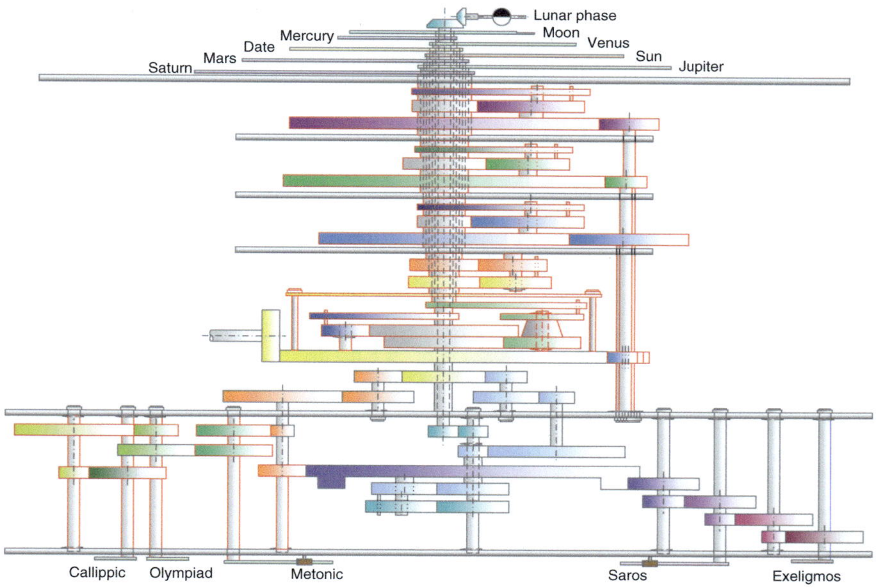

Date subsystem: Design of Fig. 5.7
Eclipse prediction subsystem: Design of Fig. 5.8
Calendrical subsystem: Design of Fig. 6.9a
Lunar subsystem: Design of Fig. 7.9a
Solar subsystem: Design of Fig. 8.9
(Inferior) Planetary subsystem: Design of Fig. 9.4a
(Superior) Planetary subsystem: Design of Fig. 9.4b
Moon phase display device: Designs of Figs. 10.6 and 10.8

Model 4

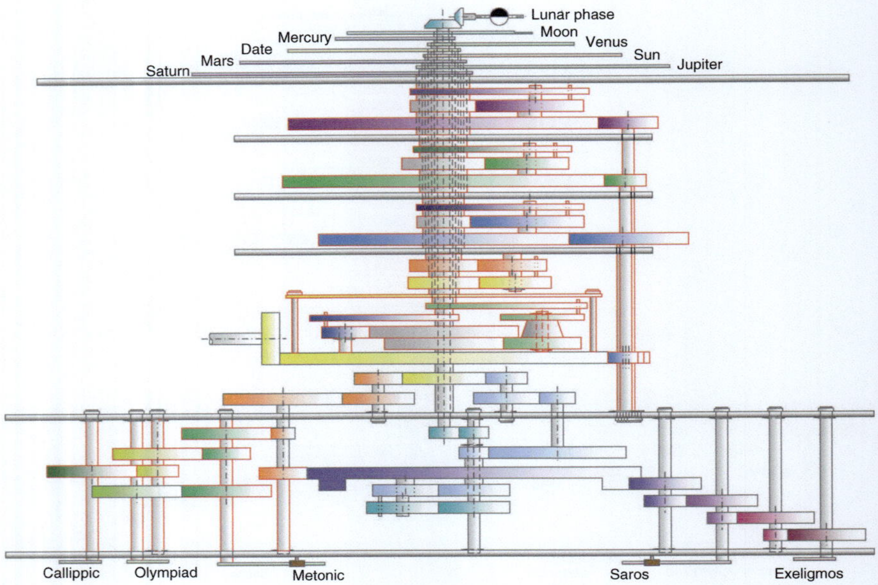

Date subsystem: Design of Fig. 5.7
Eclipse prediction subsystem: Design of Fig. 5.8
Calendrical subsystem: Design of Fig. 6.9b
Lunar subsystem: Design of Fig. 7.9a
Solar subsystem: Design of Fig. 8.9
(Inferior) Planetary subsystem: Design of Fig. 8.9a
(Superior) Planetary subsystem: Design of Fig. 9.4b
Moon phase display device: Designs of Figs. 10.6 and 10.8

Model 5

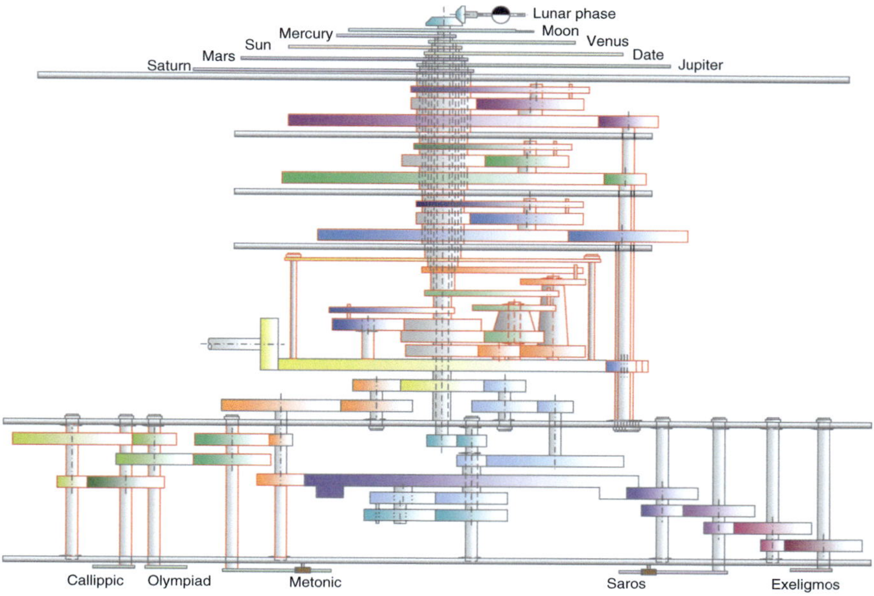

Date subsystem: Design of Fig. 5.7
Eclipse prediction subsystem: Design of Fig. 5.8
Calendrical subsystem: Design of Fig. 6.9a
Lunar subsystem: Design of Fig. 7.9a
Solar subsystem: Design of Fig. 8.14a
(Inferior) Planetary subsystem: Design of Fig. 9.4a
(Superior) Planetary subsystem: Design of Fig. 9.4b
Moon phase display device: Designs of Figs. 10.6 and 10.8

Model 6

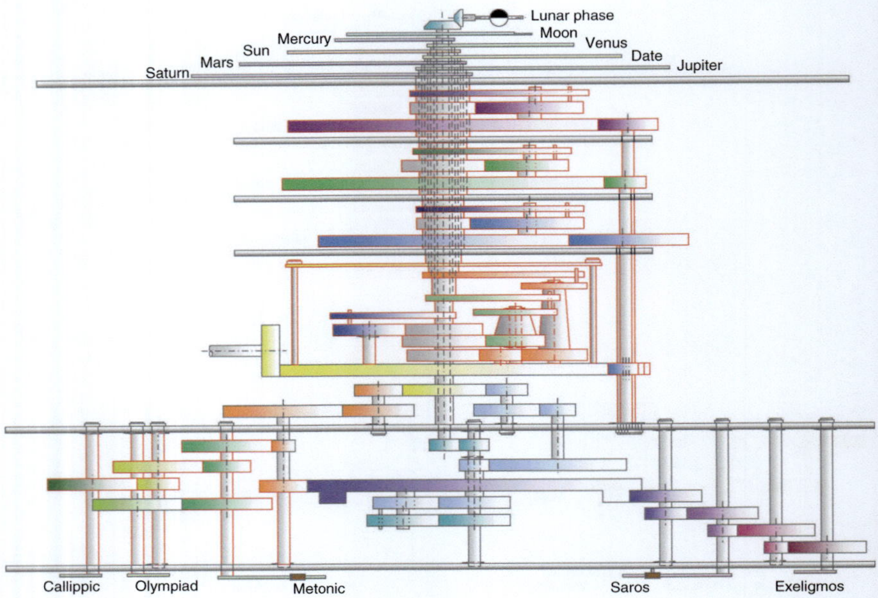

Date subsystem: Design of Fig. 5.7
Eclipse prediction subsystem: Design of Fig. 5.8
Calendrical subsystem: Design of Fig. 6.9b
Lunar subsystem: Design of Fig. 7.9a
Solar subsystem: Design of Fig. 8.14a
(Inferior) Planetary subsystem: Design of Fig. 9.4a
(Superior) Planetary subsystem: Design of Fig. 9.4b
Moon phase display device: Designs of Figs. 10.6 and 10.8

Model 7

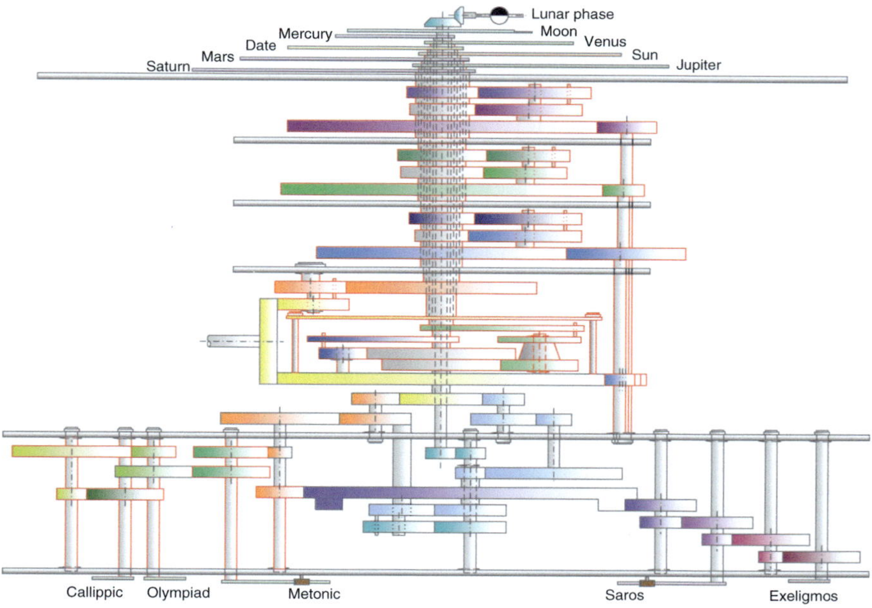

Date subsystem: Design of Fig. 5.7
Eclipse prediction subsystem: Design of Fig. 5.8
Calendrical subsystem: Design of Fig. 6.9a
Lunar subsystem: Design of Fig. 7.9a
Solar subsystem: Design of Fig. 8.5b
(Inferior) Planetary subsystem: Design of Fig. 9.4a
(Superior) Planetary subsystem: Design of Fig. 9.7a
Moon phase display device: Designs of Figs. 10.6 and 10.8

Model 8

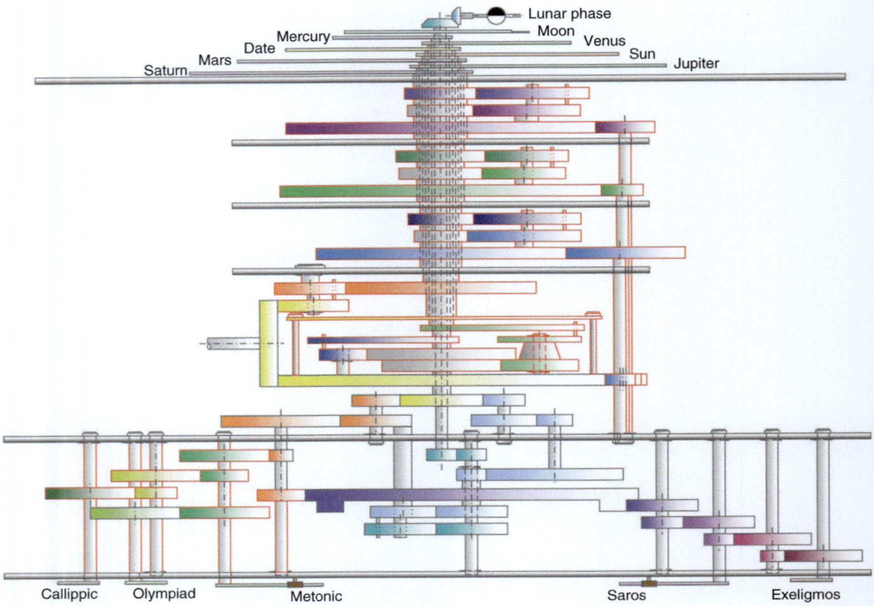

Date subsystem: Design of Fig. 5.7
Eclipse prediction subsystem: Design of Fig. 5.8
Calendrical subsystem: Design of Fig. 6.9b
Lunar subsystem: Design of Fig. 7.9a
Solar subsystem: Design of Fig. 8.5b
(Inferior) Planetary subsystem: Design of Fig. 9.4a
(Superior) Planetary subsystem: Design of Fig. 9.7a
Moon phase display device: Designs of Figs. 10.6 and 10.8

Model 9

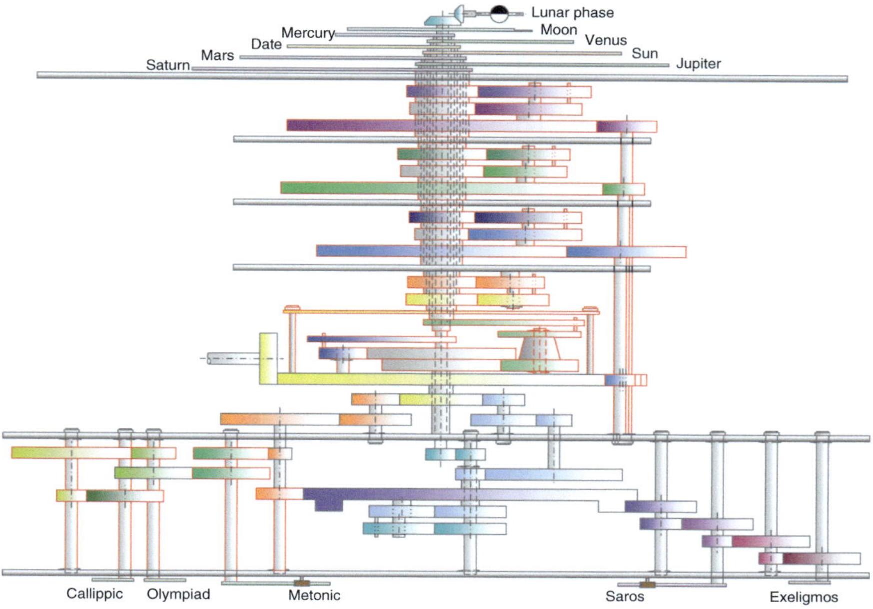

Date subsystem: Design of Fig. 5.7
Eclipse prediction subsystem: Design of Fig. 5.8
Calendrical subsystem: Design of Fig. 6.9a
Lunar subsystem: Design of Fig. 7.9a
Solar subsystem: Design of Fig. 8.9
(Inferior) Planetary subsystem: Design of Fig. 9.4a
(Superior) Planetary subsystem: Design of Fig. 9.7a
Moon phase display device: Designs of Figs. 10.6 and 10.8

Model 10

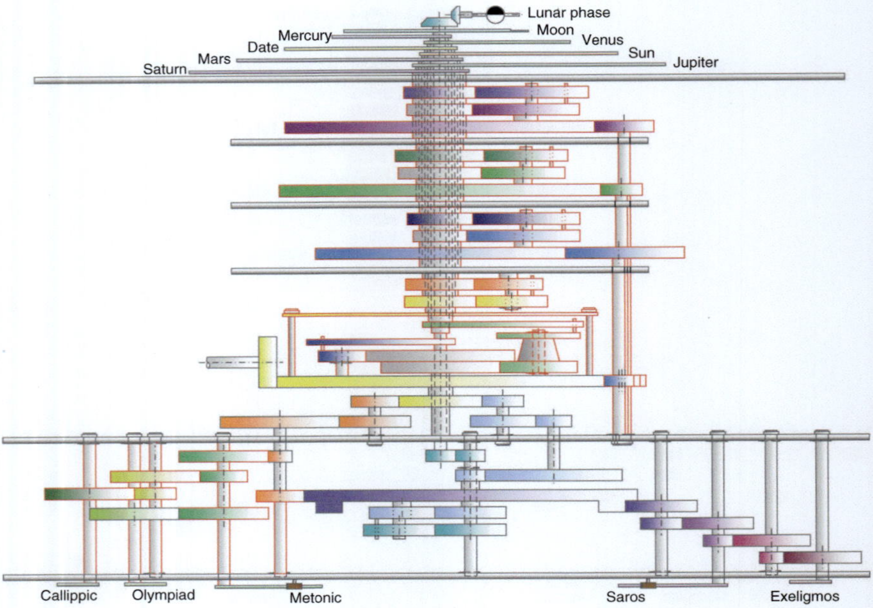

Date subsystem: Design of Fig. 5.7
Eclipse prediction subsystem: Design of Fig. 5.8
Calendrical subsystem: Design of Fig. 6.9b
Lunar subsystem: Design of Fig. 7.9a
Solar subsystem: Design of Fig. 8.9
(Inferior) Planetary subsystem: Design of Fig. 9.4a
(Superior) Planetary subsystem: Design of Fig. 9.7a
Moon phase display device: Designs of Figs. 10.6 and 10.8

Model 11

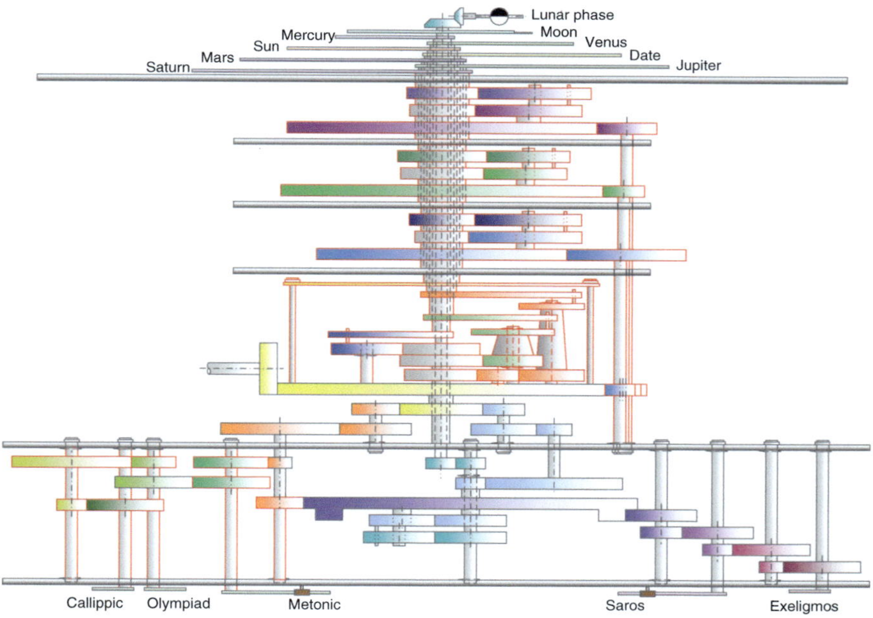

Date subsystem: Design of Fig. 5.7
Eclipse prediction subsystem: Design of Fig. 5.8
Calendrical subsystem: Design of Fig. 6.9a
Lunar subsystem: Design of Fig. 7.9a
Solar subsystem: Design of Fig. 8.14a
(Inferior) Planetary subsystem: Design of Fig. 9.4a
(Superior) Planetary subsystem: Design of Fig. 9.7a
Moon phase display device: Designs of Figs. 10.6 and 10.8

Model 12

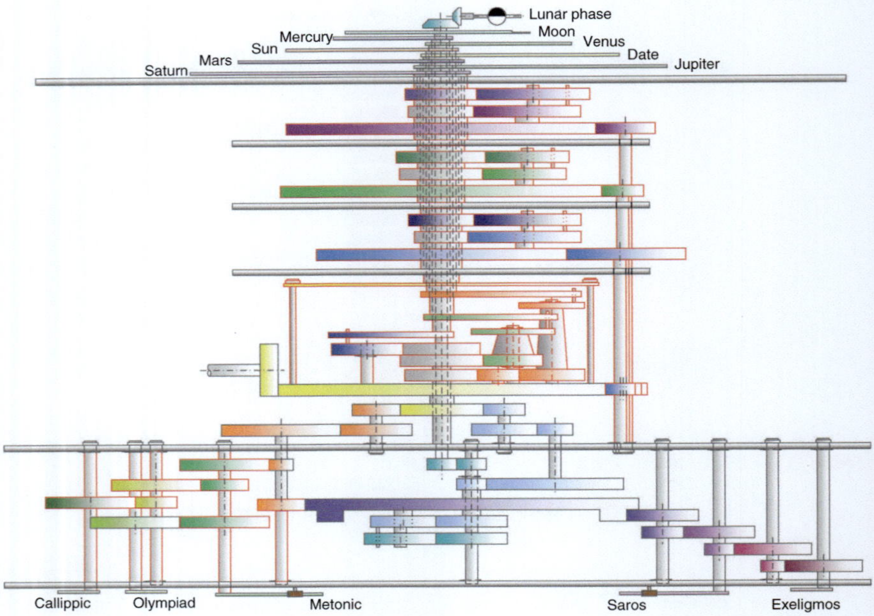

Date subsystem: Design of Fig. 5.7
Eclipse prediction subsystem: Design of Fig. 5.8
Calendrical subsystem: Design of Fig. 6.9b
Lunar subsystem: Design of Fig. 7.9a
Solar subsystem: Design of Fig. 8.14a
(Inferior) Planetary subsystem: Design of Fig. 9.4a
(Superior) Planetary subsystem: Design of Fig. 9.7a
Moon phase display device: Designs of Figs. 10.6 and 10.8

Model 13

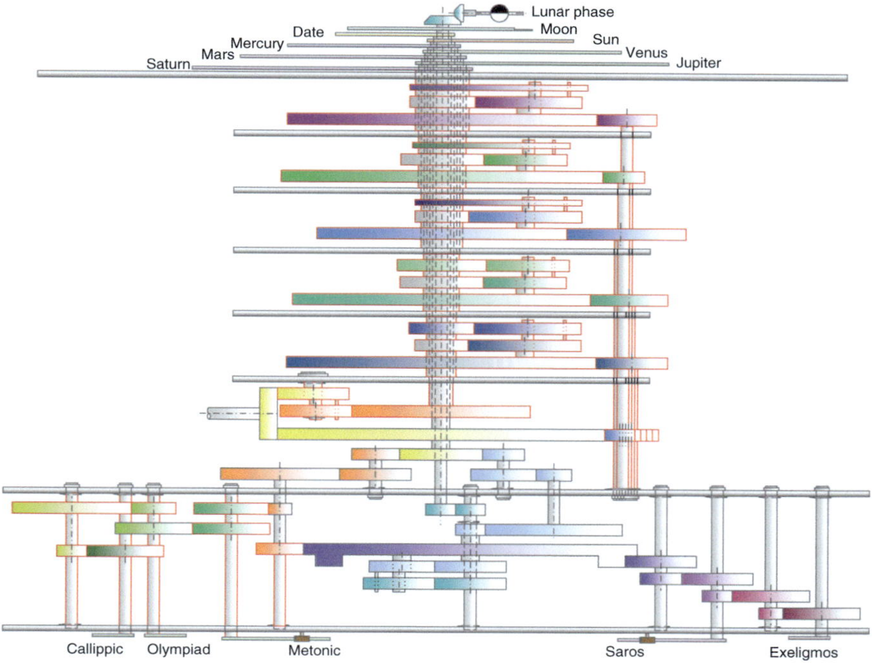

Date subsystem: Design of Fig. 5.7
Eclipse prediction subsystem: Design of Fig. 5.8
Calendrical subsystem: Design of Fig. 6.9a
Lunar subsystem: Design of Fig. 7.9a
Solar subsystem: Design of Fig. 8.5b
(Inferior) Planetary subsystem: Design of Fig. 9.7a
(Superior) Planetary subsystem: Design of Fig. 9.4b
Moon phase display device: Designs of Figs. 10.6, 10.8, and 10.10

Model 14

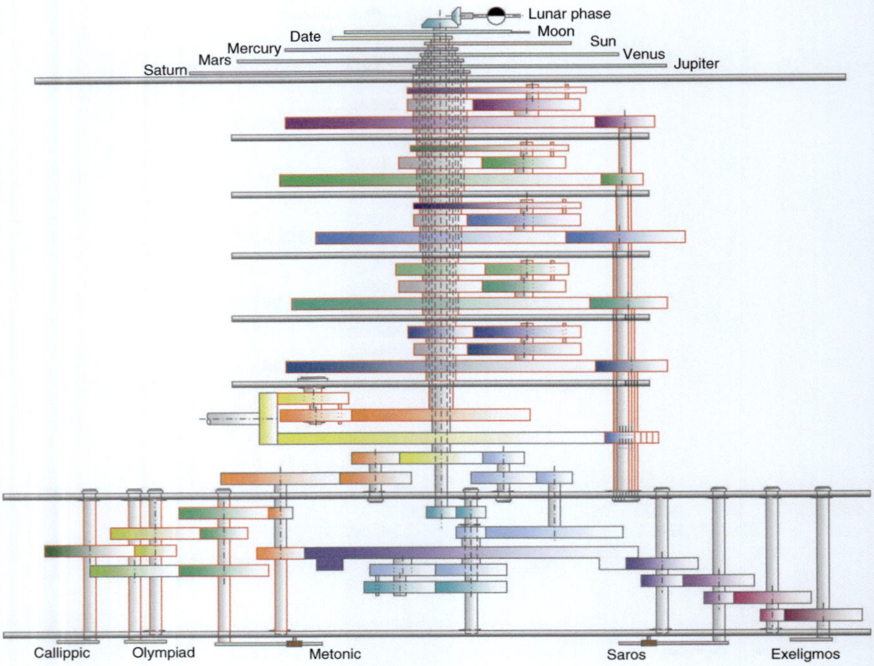

Date subsystem: Design of Fig. 5.7
Eclipse prediction subsystem: Design of Fig. 5.8
Calendrical subsystem: Design of Fig. 6.9b
Lunar subsystem: Design of Fig. 7.9a
Solar subsystem: Design of Fig. 8.5b
(Inferior) Planetary subsystem: Design of Fig. 9.7a
(Superior) Planetary subsystem: Design of Fig. 9.4b
Moon phase display device: Designs of Figs. 10.6, 10.8, and 10.10

Model 15

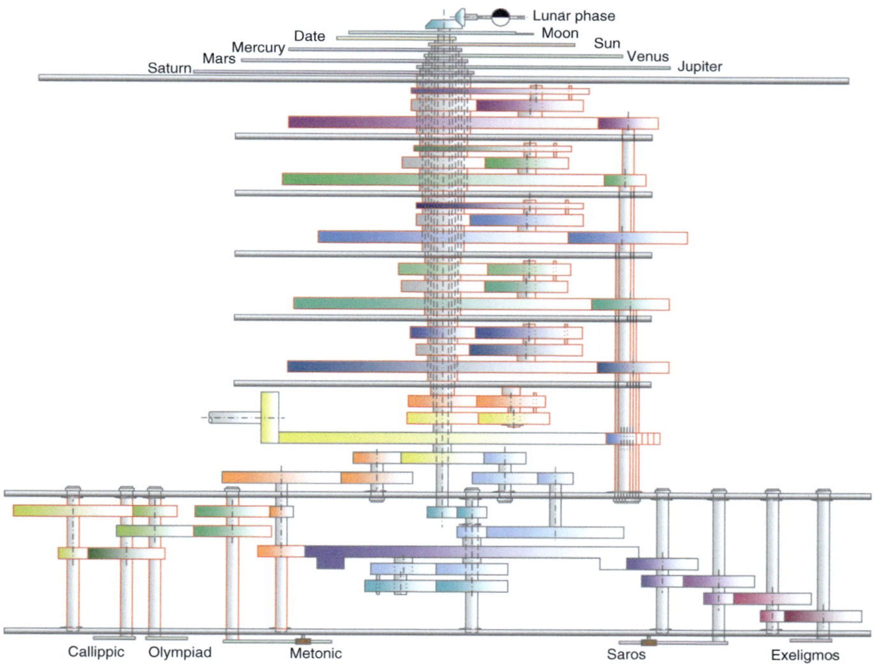

Date subsystem: Design of Fig. 5.7
Eclipse prediction subsystem: Design of Fig. 5.8
Calendrical subsystem: Design of Fig. 6.9a
Lunar subsystem: Design of Fig. 7.9a
Solar subsystem: Design of Fig. 8.9
(Inferior) Planetary subsystem: Design of Fig. 9.7a
(Superior) Planetary subsystem: Design of Fig. 9.4b
Moon phase display device: Designs of Figs. 10.6, 10.8, and 10.10

Model 16

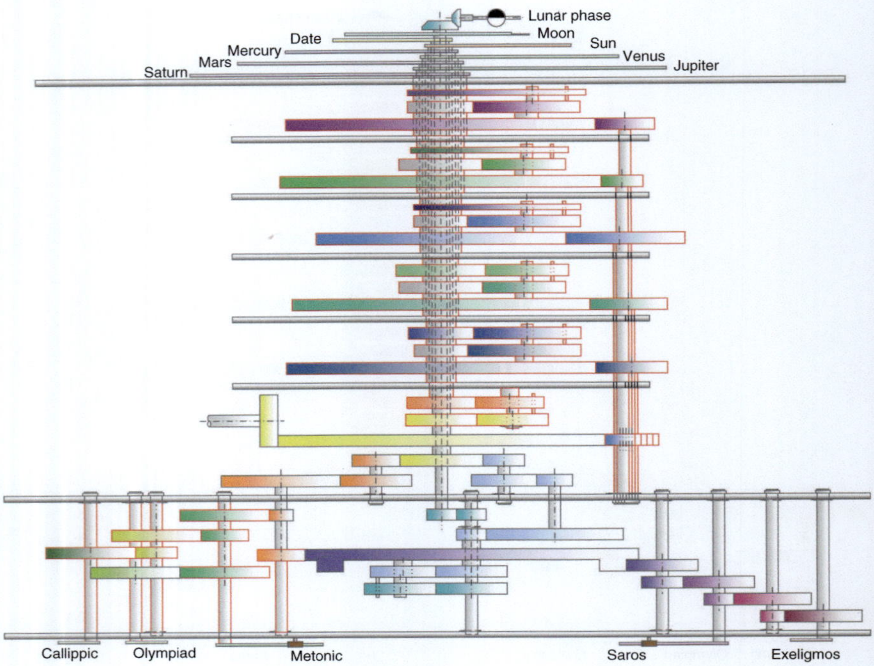

Date subsystem: Design of Fig. 5.7
Eclipse prediction subsystem: Design of Fig. 5.8
Calendrical subsystem: Design of Fig. 6.9b
Lunar subsystem: Design of Fig. 7.9a
Solar subsystem: Design of Fig. 8.9
(Inferior) Planetary subsystem: Design of Fig. 9.7a
(Superior) Planetary subsystem: Design of Fig. 9.4b
Moon phase display device: Designs of Figs. 10.6, 10.8, and 10.10

Model 17

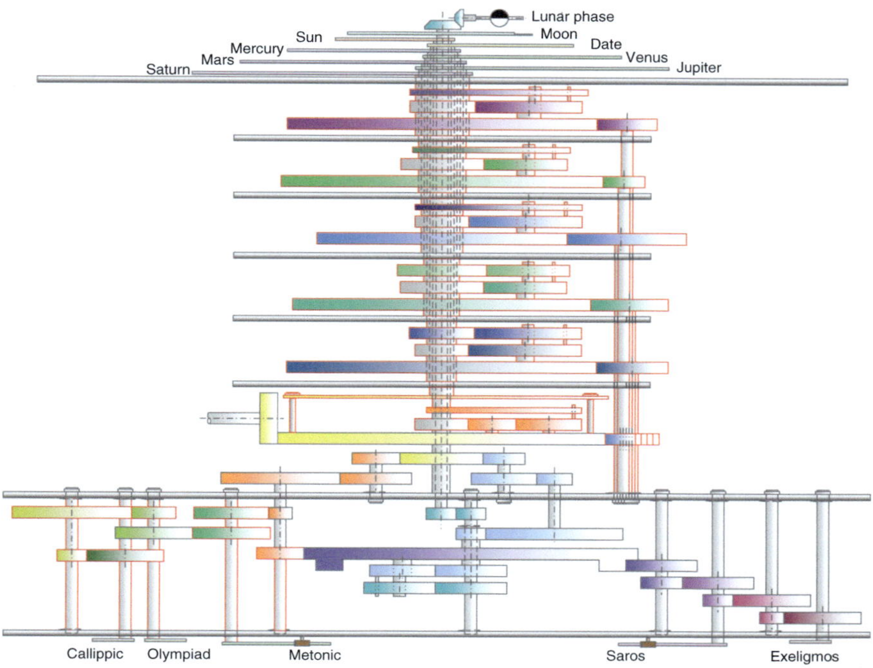

Date subsystem: Design of Fig. 5.7
Eclipse prediction subsystem: Design of Fig. 5.8
Calendrical subsystem: Design of Fig. 6.9a
Lunar subsystem: Design of Fig. 7.9a
Solar subsystem: Design of Fig. 8.14a
(Inferior) Planetary subsystem: Design of Fig. 9.7a
(Superior) Planetary subsystem: Design of Fig. 9.4b
Moon phase display device: Designs of Figs. 10.6, 10.8, and 10.10

Model 18

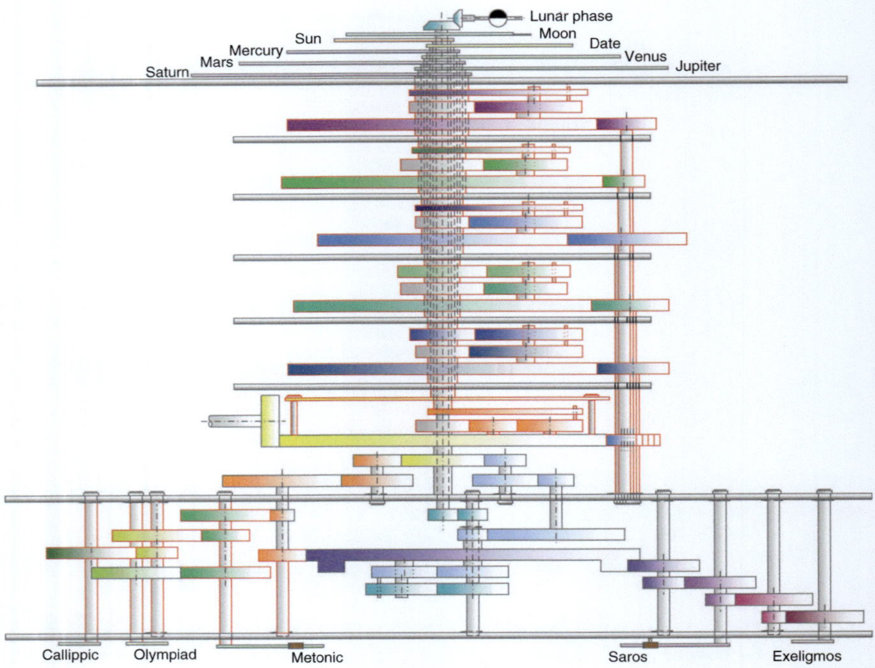

Date subsystem: Design of Fig. 5.7
Eclipse prediction subsystem: Design of Fig. 5.8
Calendrical subsystem: Design of Fig. 6.9b
Lunar subsystem: Design of Fig. 7.9a
Solar subsystem: Design of Fig. 8.14a
(Inferior) Planetary subsystem: Design of Fig. 9.7a
(Superior) Planetary subsystem: Design of Fig. 9.4b
Moon phase display device: Designs of Figs. 10.6, 10.8, and 10.10

Model 19

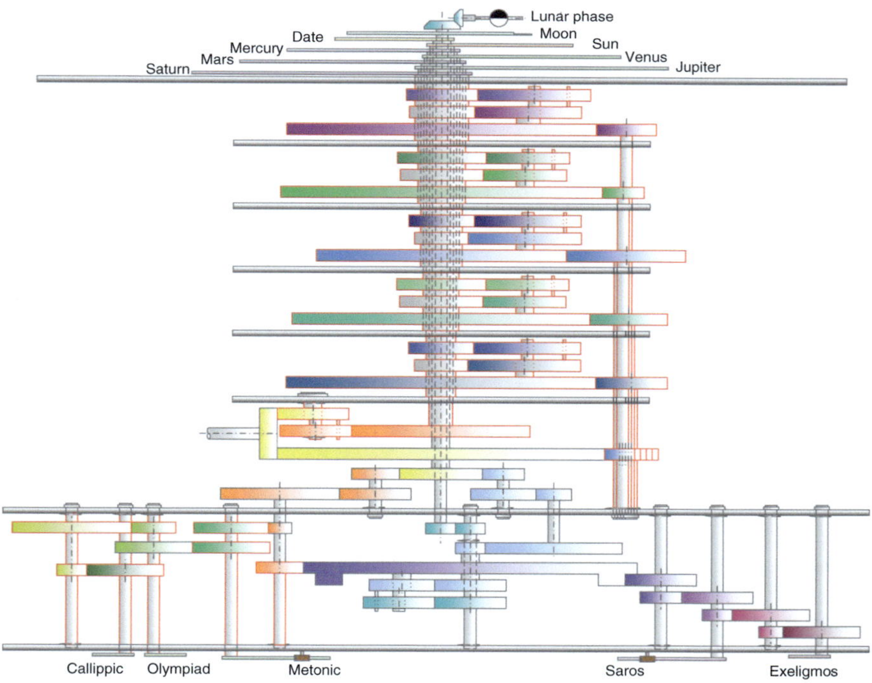

Date subsystem: Design of Fig. 5.7
Eclipse prediction subsystem: Design of Fig. 5.8
Calendrical subsystem: Design of Fig. 6.9a
Lunar subsystem: Design of Fig. 7.9a
Solar subsystem: Design of Fig. 8.5b
(Inferior) Planetary subsystem: Design of Fig. 9.7a
(Superior) Planetary subsystem: Design of Fig. 9.7a
Moon phase display device: Designs of Figs. 10.6, 10.8, and 10.10

Model 20

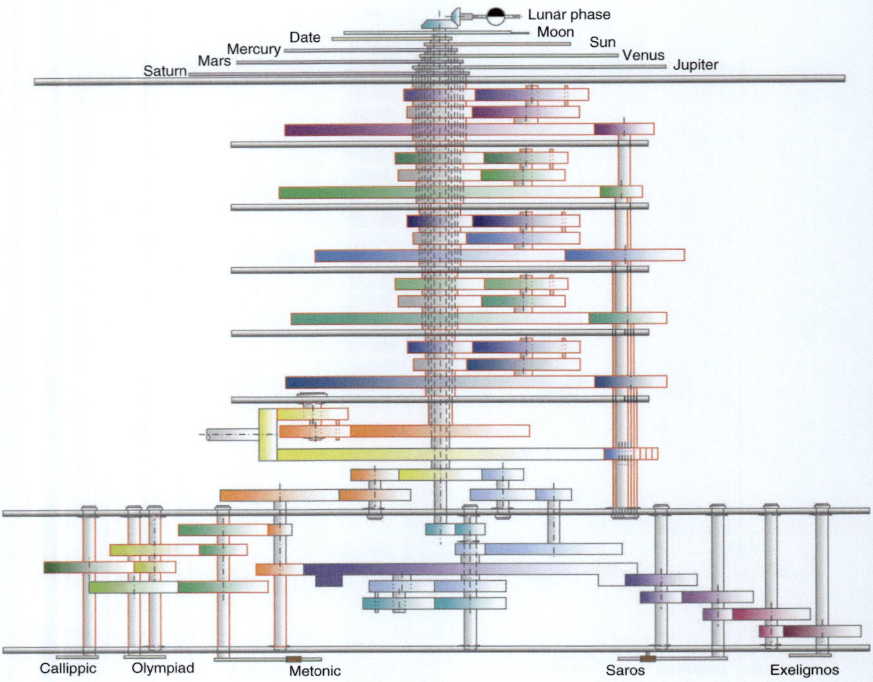

Date subsystem: Design of Fig. 5.7
Eclipse prediction subsystem: Design of Fig. 5.8
Calendrical subsystem: Design of Fig. 6.9b
Lunar subsystem: Design of Fig. 7.9a
Solar subsystem: Design of Fig. 8.5b
(Inferior) Planetary subsystem: Design of Fig. 9.7a
(Superior) Planetary subsystem: Design of Fig. 9.7a
Moon phase display device: Designs of Figs. 10.6, 10.8, and 10.10

Model 21

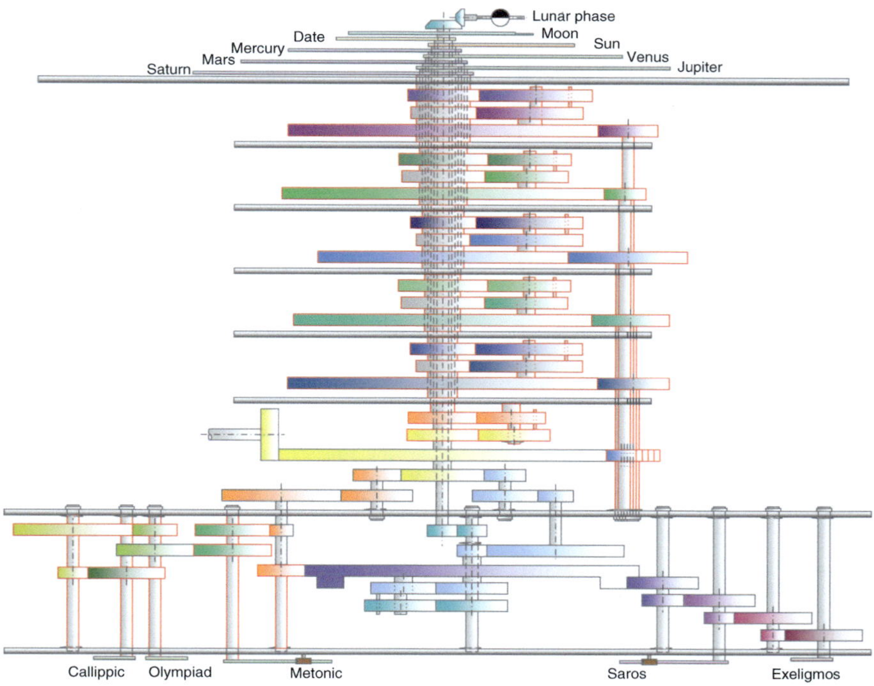

Date subsystem: Design of Fig. 5.7
Eclipse prediction subsystem: Design of Fig. 5.8
Calendrical subsystem: Design of Fig. 6.9a
Lunar subsystem: Design of Fig. 7.9a
Solar subsystem: Design of Fig. 8.9
(Inferior) Planetary subsystem: Design of Fig. 9.7a
(Superior) Planetary subsystem: Design of Fig. 9.7a
Moon phase display device: Designs of Figs. 10.6, 10.8, and 10.10

Model 22

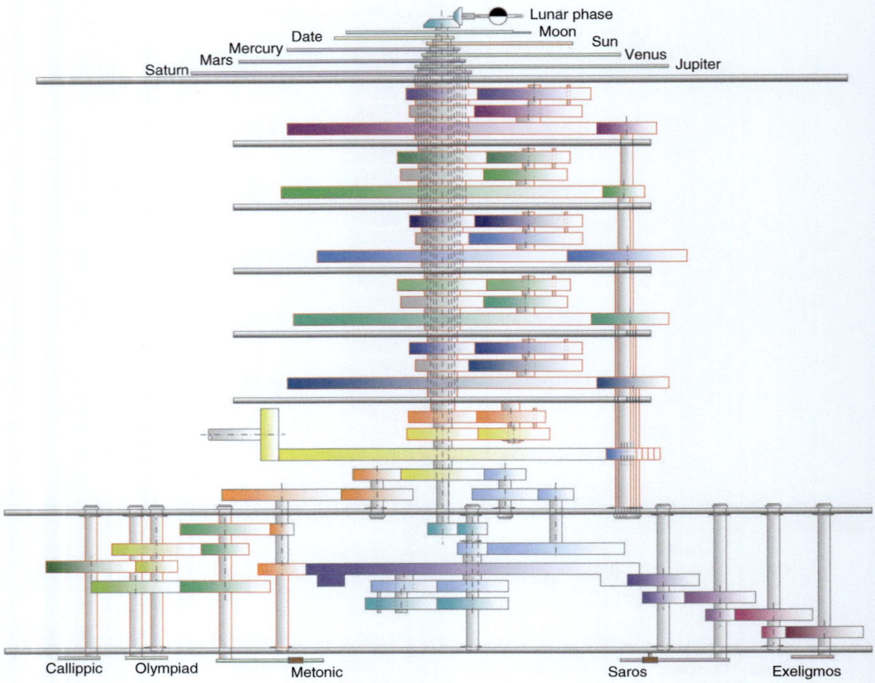

Date subsystem: Design of Fig. 5.7
Eclipse prediction subsystem: Design of Fig. 5.8
Calendrical subsystem: Design of Fig. 6.9b
Lunar subsystem: Design of Fig. 7.9a
Solar subsystem: Design of Fig. 8.9
(Inferior) Planetary subsystem: Design of Fig. 9.7a
(Superior) Planetary subsystem: Design of Fig. 9.7a
Moon phase display device: Designs of Figs. 10.6, 10.8, and 10.10

Model 23

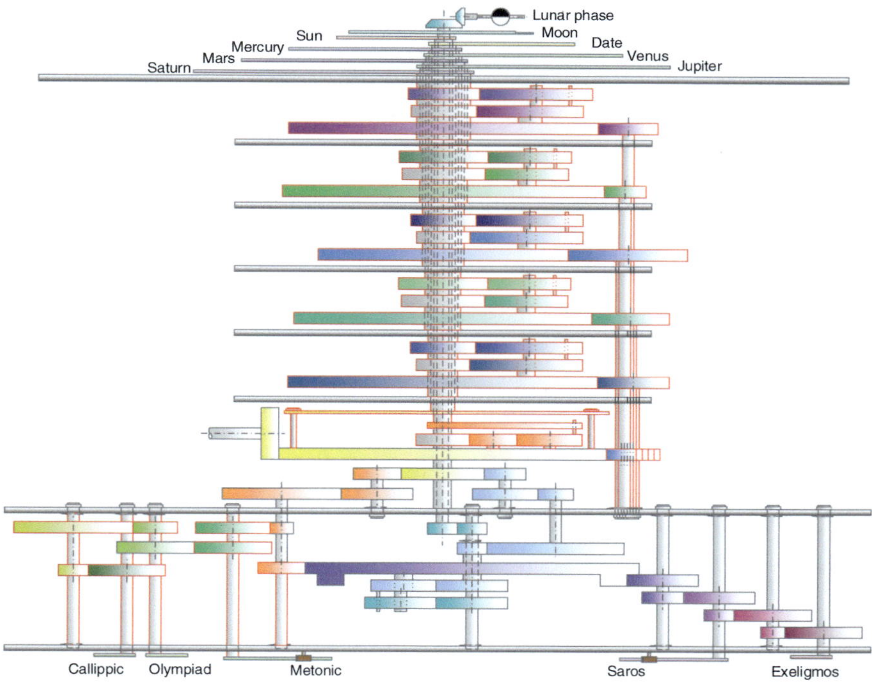

Date subsystem: Design of Fig. 5.7
Eclipse prediction subsystem: Design of Fig. 5.8
Calendrical subsystem: Design of Fig. 6.9a
Lunar subsystem: Design of Fig. 7.9a
Solar subsystem: Design of Fig. 8.14a
(Inferior) Planetary subsystem: Design of Fig. 9.7a
(Superior) Planetary subsystem: Design of Fig. 9.7a
Moon phase display device: Designs of Figs. 10.6, 10.8, and 10.10

Model 24

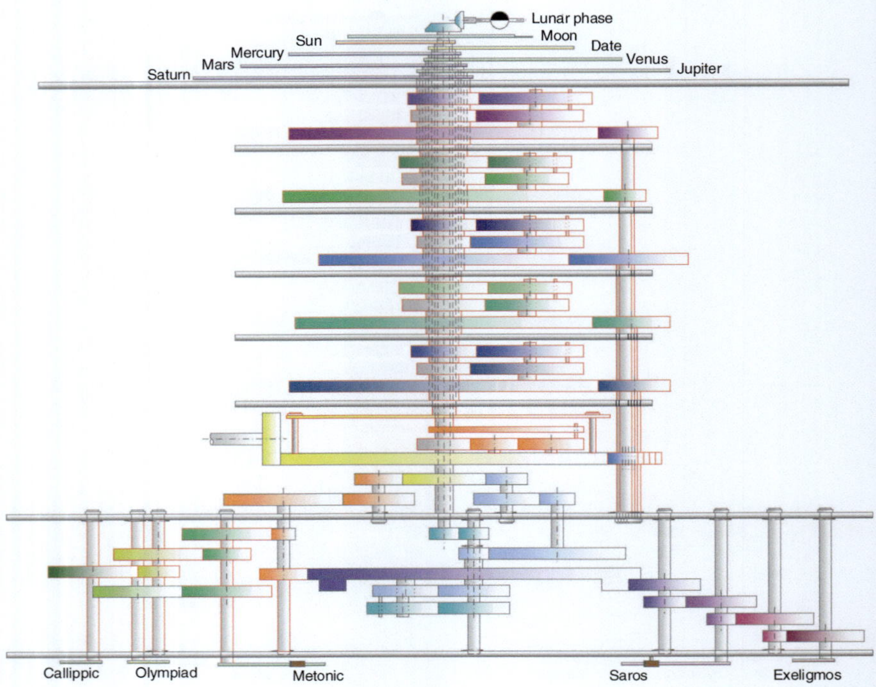

Date subsystem: Design of Fig. 5.7
Eclipse prediction subsystem: Design of Fig. 5.8
Calendrical subsystem: Design of Fig. 6.9b
Lunar subsystem: Design of Fig. 7.9a
Solar subsystem: Design of Fig. 8.14a
(Inferior) Planetary subsystem: Design of Fig. 9.7a
(Superior) Planetary subsystem: Design of Fig. 9.7a
Moon phase display device: Designs of Figs. 10.6, 10.8, and 10.10

Model 25

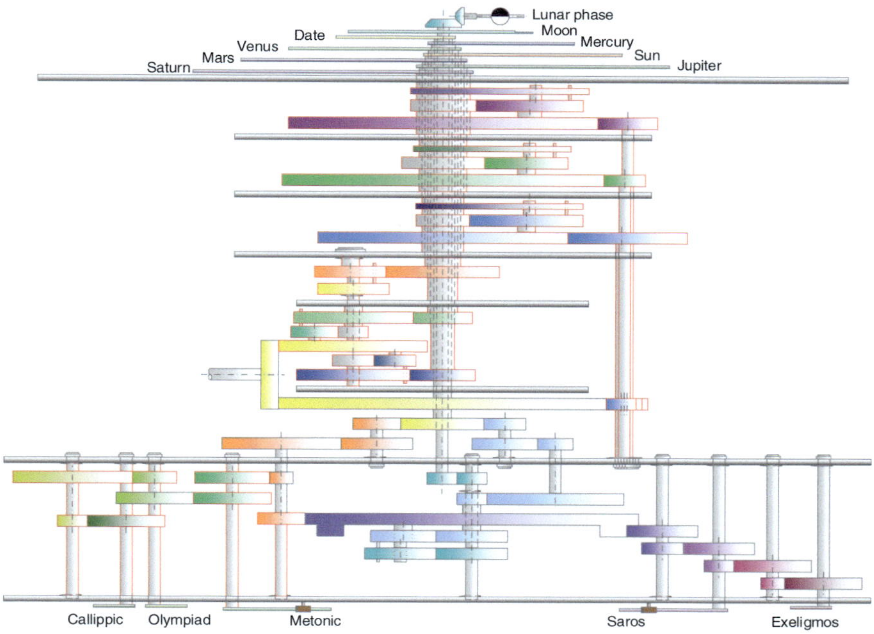

Date subsystem: Design of Fig. 5.7
Eclipse prediction subsystem: Design of Fig. 5.8
Calendrical subsystem: Design of Fig. 6.9a
Lunar subsystem: Design of Fig. 7.9a
Solar subsystem: Design of Fig. 8.5b
(Inferior) Planetary subsystem: Design of Fig. 9.7b
(Superior) Planetary subsystem: Design of Fig. 9.4b
Moon phase display device: Designs of Figs. 10.6, 10.8, and 10.10

Model 26

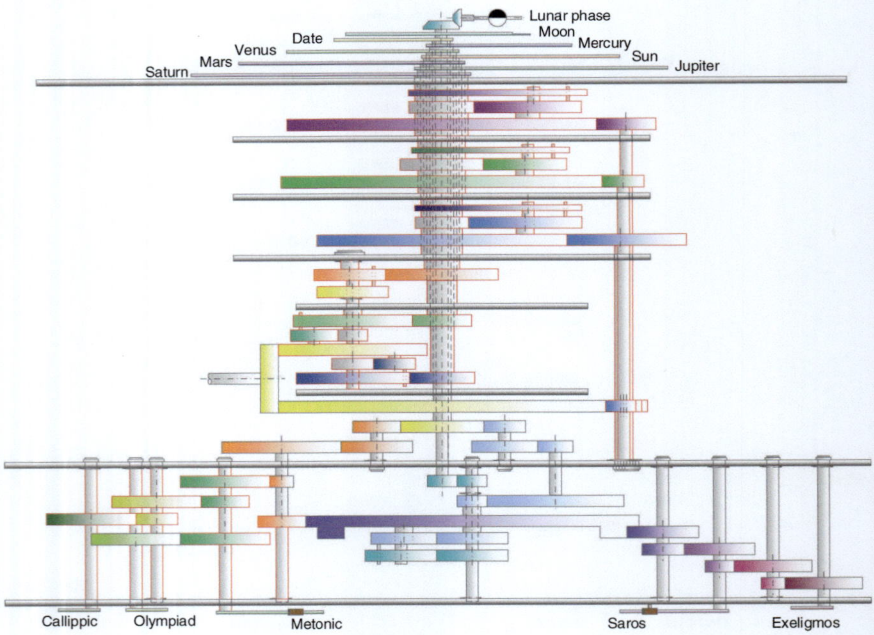

Date subsystem: Design of Fig. 5.7
Eclipse prediction subsystem: Design of Fig. 5.8
Calendrical subsystem: Design of Fig. 6.9b
Lunar subsystem: Design of Fig. 7.9a
Solar subsystem: Design of Fig. 8.5b
(Inferior) Planetary subsystem: Design of Fig. 9.7b
(Superior) Planetary subsystem: Design of Fig. 9.4b
Moon phase display device: Designs of Figs. 10.6, 10.8, and 10.10

Model 27

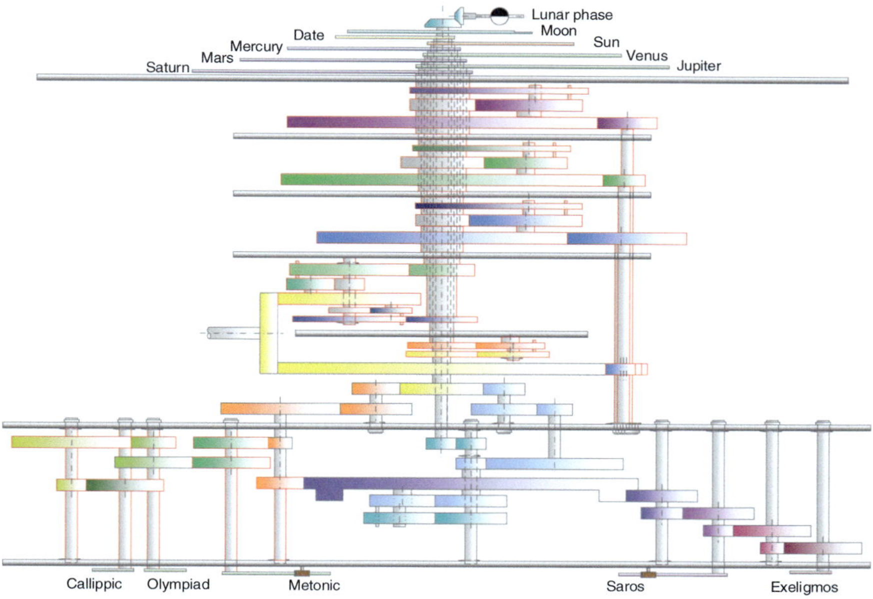

Date subsystem: Design of Fig. 5.7
Eclipse prediction subsystem: Design of Fig. 5.8
Calendrical subsystem: Design of Fig. 6.9a
Lunar subsystem: Design of Fig. 7.9a
Solar subsystem: Design of Fig. 8.9
(Inferior) Planetary subsystem: Design of Fig. 9.7b
(Superior) Planetary subsystem: Design of Fig. 9.4b
Moon phase display device: Designs of Figs. 10.6, 10.8, and 10.10

Model 28

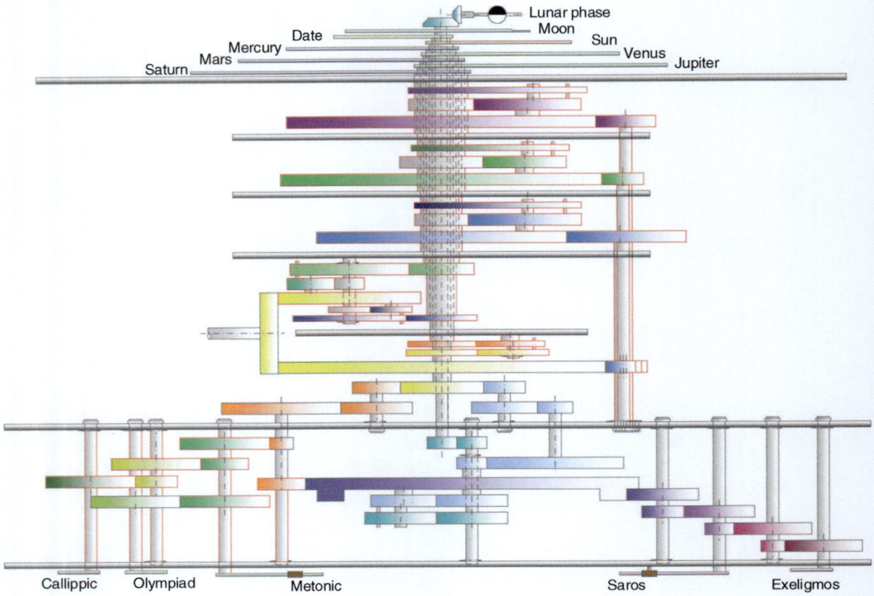

Date subsystem: Design of Fig. 5.7
Eclipse prediction subsystem: Design of Fig. 5.8
Calendrical subsystem: Design of Fig. 6.9b
Lunar subsystem: Design of Fig. 7.9a
Solar subsystem: Design of Fig. 8.9
(Inferior) Planetary subsystem: Design of Fig. 9.7b
(Superior) Planetary subsystem: Design of Fig. 9.4b
Moon phase display device: Designs of Figs. 10.6, 10.8, and 10.10

Model 29

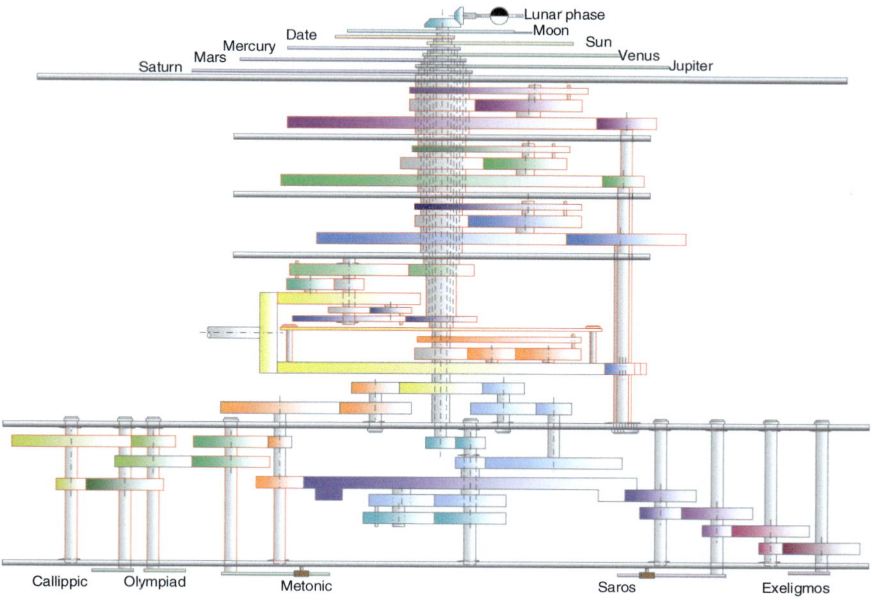

Date subsystem: Design of Fig. 5.7
Eclipse prediction subsystem: Design of Fig. 5.8
Calendrical subsystem: Design of Fig. 6.9a
Lunar subsystem: Design of Fig. 7.9a
Solar subsystem: Design of Fig. 8.14a
(Inferior) Planetary subsystem: Design of Fig. 9.7b
(Superior) Planetary subsystem: Design of Fig. 9.4b
Moon phase display device: Designs of Figs. 10.6, 10.8, and 10.10

Model 30

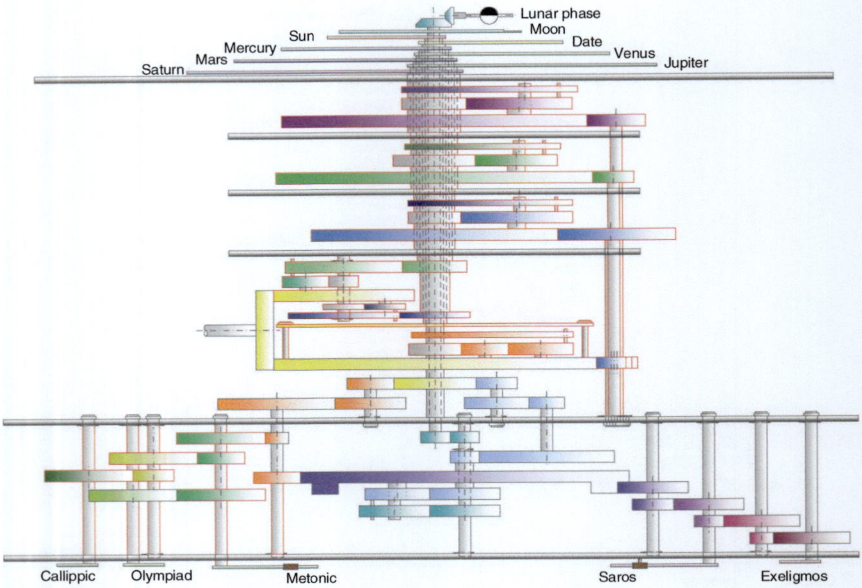

Date subsystem: Design of Fig. 5.7
Eclipse prediction subsystem: Design of Fig. 5.8
Calendrical subsystem: Design of Fig. 6.9b
Lunar subsystem: Design of Fig. 7.9a
Solar subsystem: Design of Fig. 8.14a
(Inferior) Planetary subsystem: Design of Fig. 9.7b
(Superior) Planetary subsystem: Design of Fig. 9.4b
Moon phase display device: Designs of Figs. 10.6, 10.8, and 10.10

Model 31

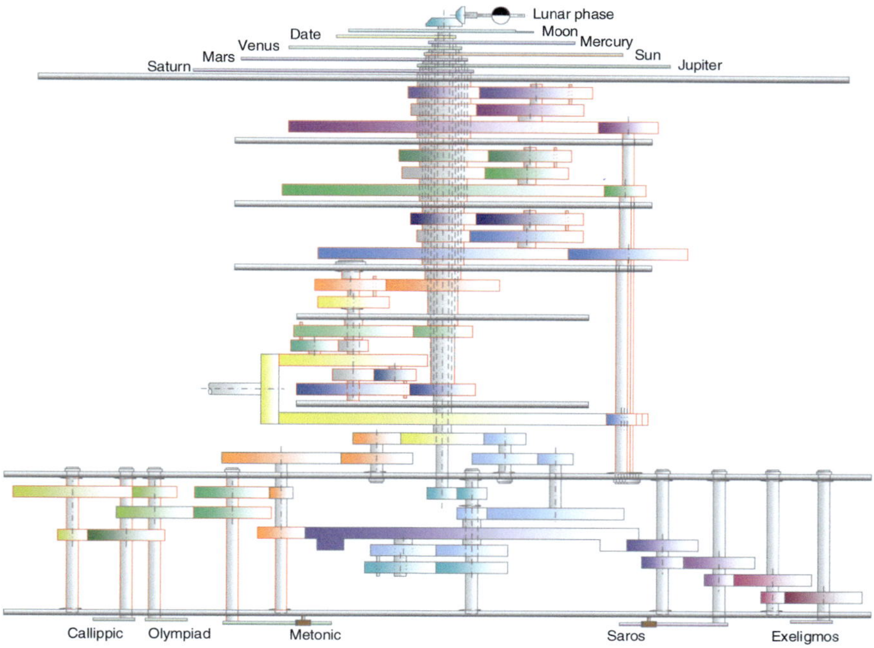

Date subsystem: Design of Fig. 5.7
Eclipse prediction subsystem: Design of Fig. 5.8
Calendrical subsystem: Design of Fig. 6.9a
Lunar subsystem: Design of Fig. 7.9a
Solar subsystem: Design of Fig. 8.5b
(Inferior) Planetary subsystem: Design of Fig. 9.7b
(Superior) Planetary subsystem: Design of Fig. 9.7a
Moon phase display device: Designs of Figs. 10.6, 10.8, and 10.10

Model 32

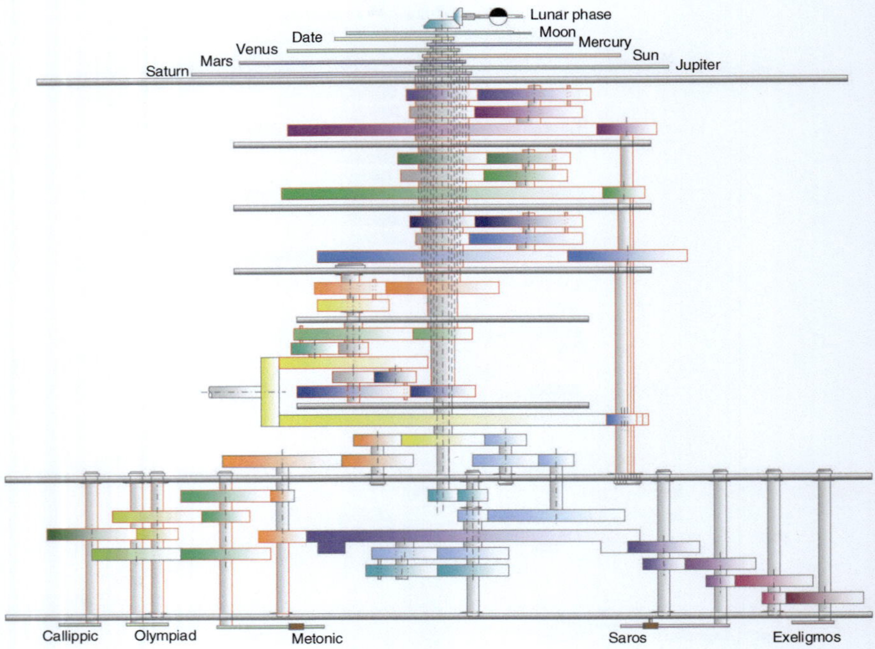

Date subsystem: Design of Fig. 5.7
Eclipse prediction subsystem: Design of Fig. 5.8
Calendrical subsystem: Design of Fig. 6.9b
Lunar subsystem: Design of Fig. 7.9a
Solar subsystem: Design of Fig. 8.5b
(Inferior) Planetary subsystem: Design of Fig. 9.7b
(Superior) Planetary subsystem: Design of Fig. 9.7a
Moon phase display device: Designs of Figs. 10.6, 10.8, and 10.10

Model 33

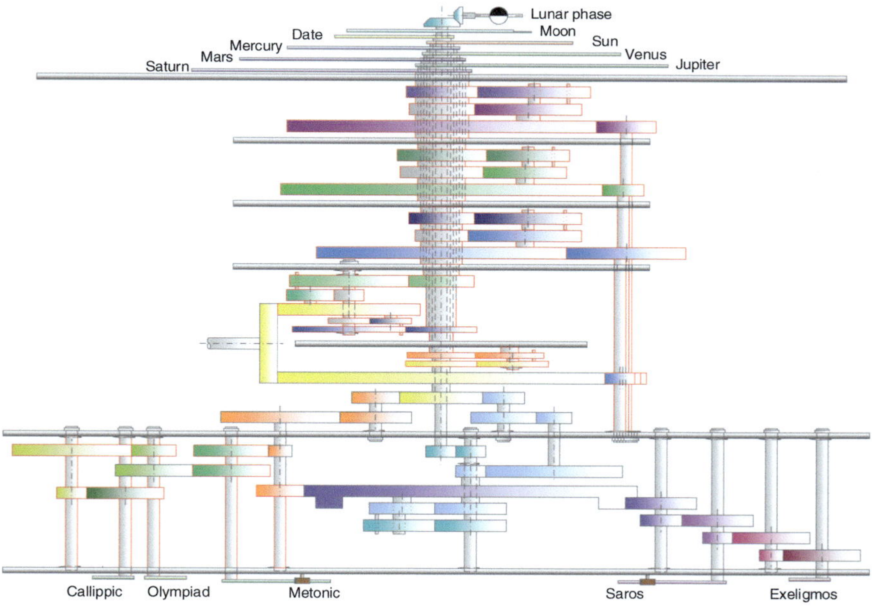

Date subsystem: Design of Fig. 5.7
Eclipse prediction subsystem: Design of Fig. 5.8
Calendrical subsystem: Design of Fig. 6.9a
Lunar subsystem: Design of Fig. 7.9a
Solar subsystem: Design of Fig. 8.9
(Inferior) Planetary subsystem: Design of Fig. 9.7b
(Superior) Planetary subsystem: Design of Fig. 9.7a
Moon phase display device: Designs of Figs. 10.6, 10.8, and 10.10

Model 34

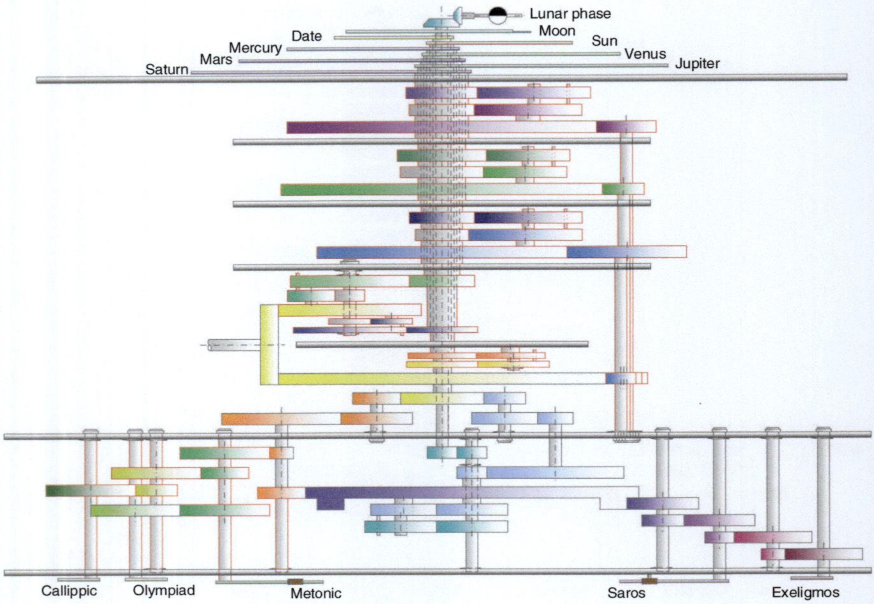

Date subsystem: Design of Fig. 5.7
Eclipse prediction subsystem: Design of Fig. 5.8
Calendrical subsystem: Design of Fig. 6.9b
Lunar subsystem: Design of Fig. 7.9a
Solar subsystem: Design of Fig. 8.9
(Inferior) Planetary subsystem: Design of Fig. 9.7b
(Superior) Planetary subsystem: Design of Fig. 9.7a
Moon phase display device: Designs of Figs. 10.6, 10.8, and 10.10

Model 35

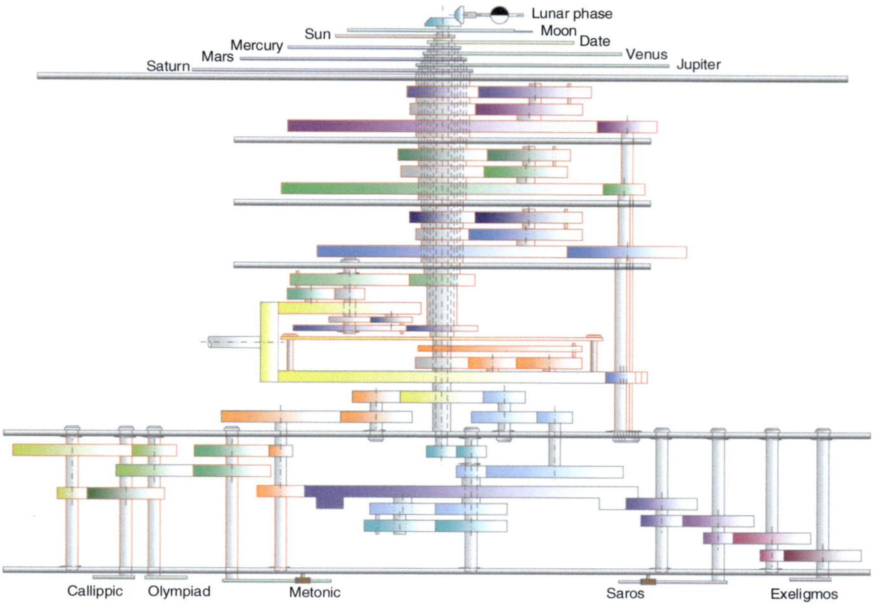

Date subsystem: Design of Fig. 5.7
Eclipse prediction subsystem: Design of Fig. 5.8
Calendrical subsystem: Design of Fig. 6.9a
Lunar subsystem: Design of Fig. 7.9a
Solar subsystem: Design of Fig. 8.14a
(Inferior) Planetary subsystem: Design of Fig. 9.7b
(Superior) Planetary subsystem: Design of Fig. 9.7a
Moon phase display device: Designs of Figs. 10.6, 10.8, and 10.10

Model 36

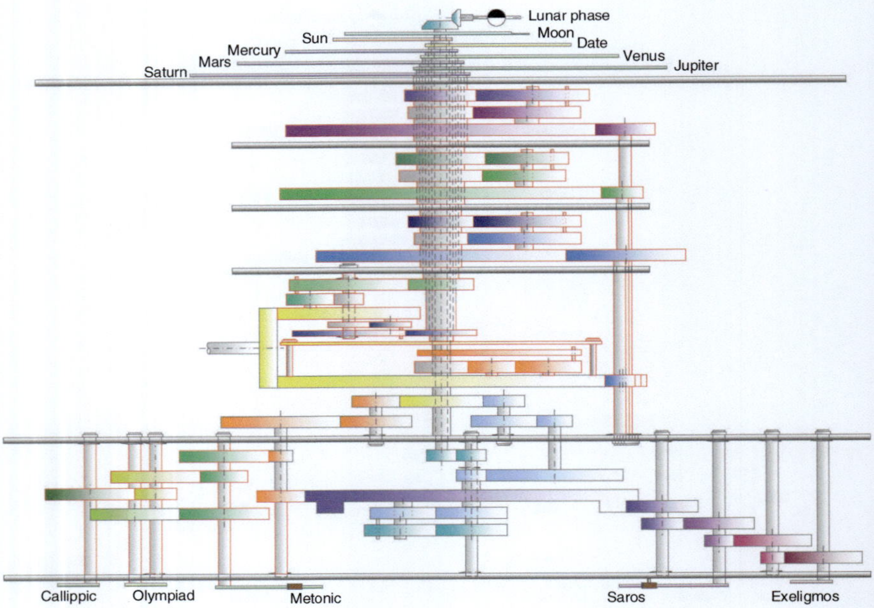

Date subsystem: Design of Fig. 5.7
Eclipse prediction subsystem: Design of Fig. 5.8
Calendrical subsystem: Design of Fig. 6.9b
Lunar subsystem: Design of Fig. 7.9a
Solar subsystem: Design of Fig. 8.14a
(Inferior) Planetary subsystem: Design of Fig. 9.7b
(Superior) Planetary subsystem: Design of Fig. 9.7a
Moon phase display device: Designs of Figs. 10.6, 10.8, and 10.10

Model 37

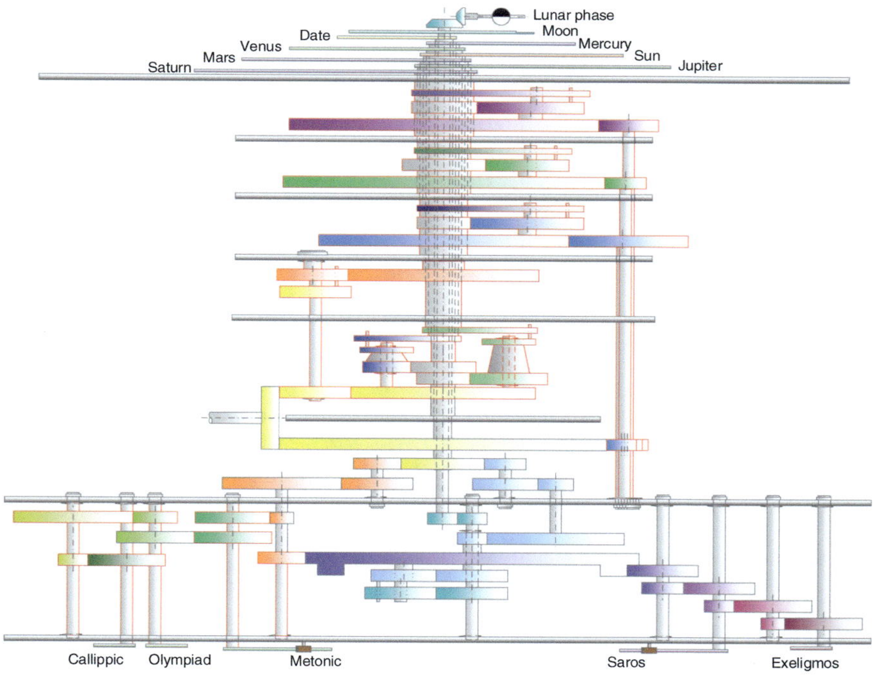

Date subsystem: Design of Fig. 5.7
Eclipse prediction subsystem: Design of Fig. 5.8
Calendrical subsystem: Design of Fig. 6.9a
Lunar subsystem: Design of Fig. 7.9a
Solar subsystem: Design of Fig. 8.5b
(Inferior) Planetary subsystem: Design of Fig. 9.7c
(Superior) Planetary subsystem: Design of Fig. 9.4b
Moon phase display device: Designs of Figs. 10.6, 10.8, and 10.10

Model 38

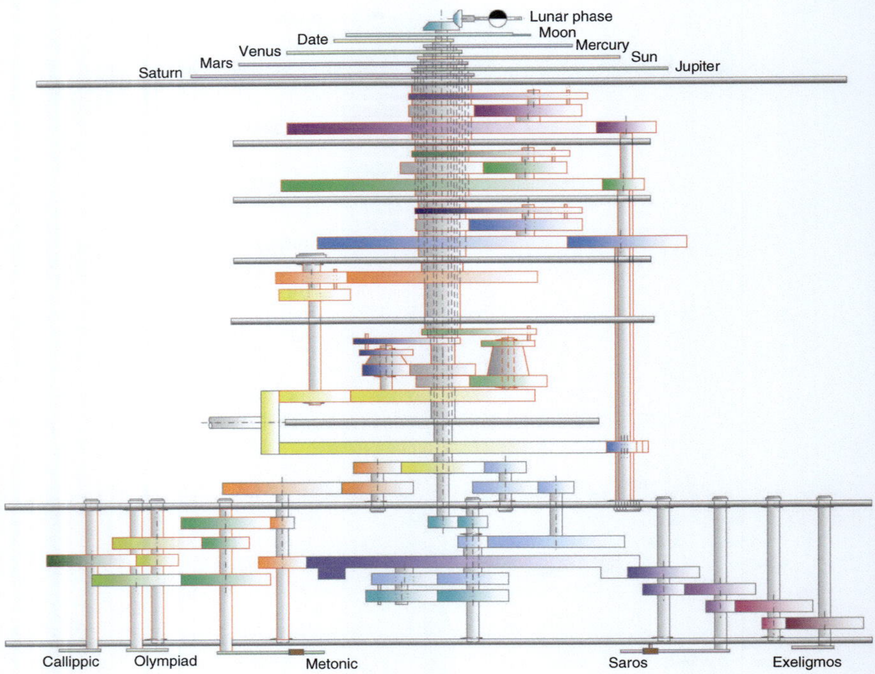

Date subsystem: Design of Fig. 5.7
Eclipse prediction subsystem: Design of Fig. 5.8
Calendrical subsystem: Design of Fig. 6.9b
Lunar subsystem: Design of Fig. 7.9a
Solar subsystem: Design of Fig. 8.5b
(Inferior) Planetary subsystem: Design of Fig. 9.7c
(Superior) Planetary subsystem: Design of Fig. 9.4b
Moon phase display device: Designs of Figs. 10.6, 10.8, and 10.10

Model 39

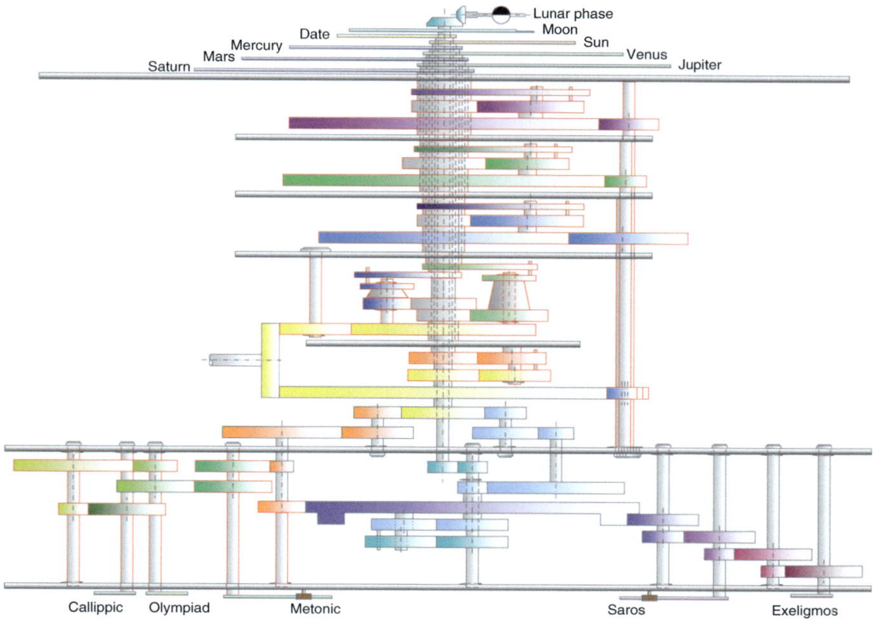

Date subsystem: Design of Fig. 5.7
Eclipse prediction subsystem: Design of Fig. 5.8
Calendrical subsystem: Design of Fig. 6.9a
Lunar subsystem: Design of Fig. 7.9a
Solar subsystem: Design of Fig. 8.9
(Inferior) Planetary subsystem: Design of Fig. 9.7c
(Superior) Planetary subsystem: Design of Fig. 9.4b
Moon phase display device: Designs of Figs. 10.6, 10.8, and 10.10

Model 40

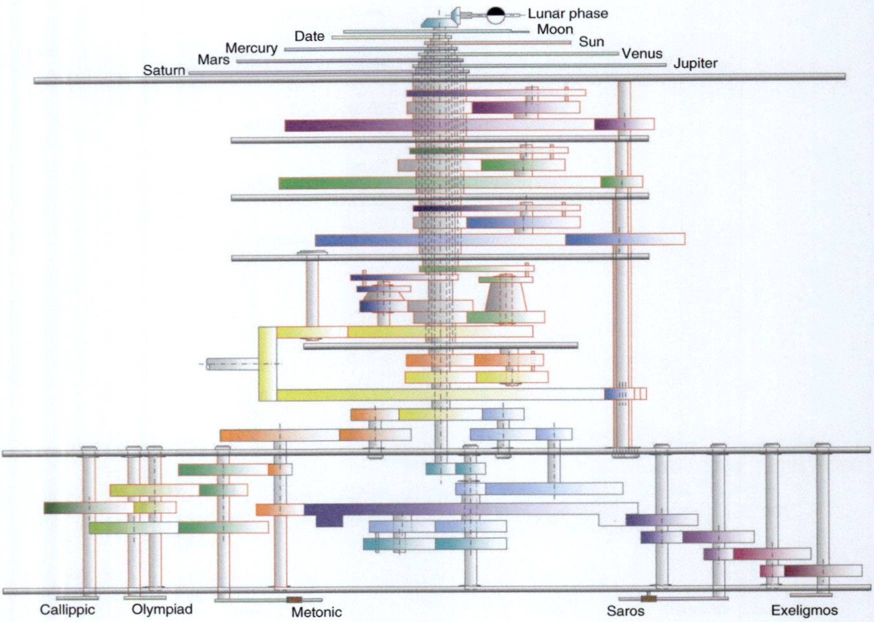

Date subsystem: Design of Fig. 5.7
Eclipse prediction subsystem: Design of Fig. 5.8
Calendrical subsystem: Design of Fig. 6.9b
Lunar subsystem: Design of Fig. 7.9a
Solar subsystem: Design of Fig. 8.9
(Inferior) Planetary subsystem: Design of Fig. 9.7c
(Superior) Planetary subsystem: Design of Fig. 9.4b
Moon phase display device: Designs of Figs. 10.6, 10.8, and 10.10

Model 41

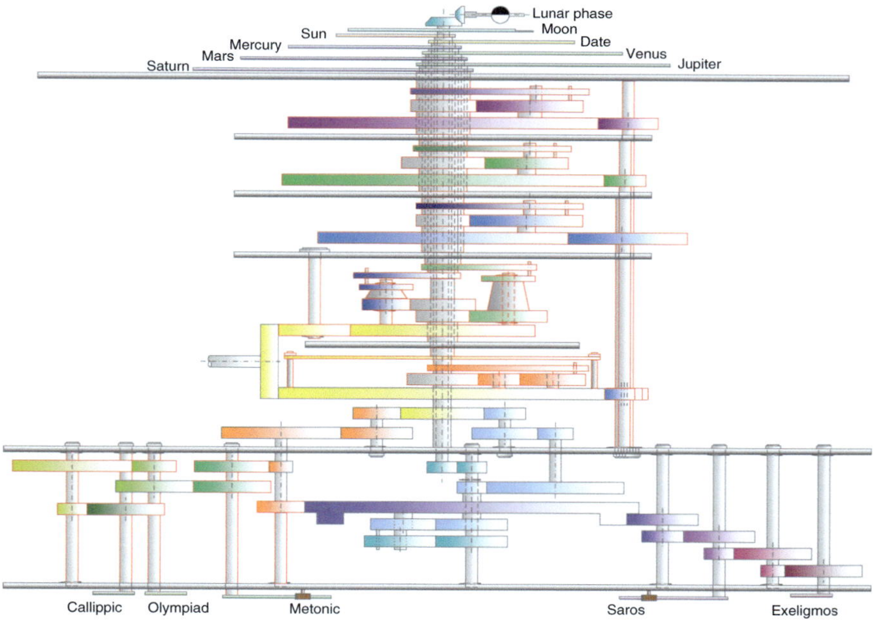

Date subsystem: Design of Fig. 5.7
Eclipse prediction subsystem: Design of Fig. 5.8
Calendrical subsystem: Design of Fig. 6.9a
Lunar subsystem: Design of Fig. 7.9a
Solar subsystem: Design of Fig. 8.14a
(Inferior) Planetary subsystem: Design of Fig. 9.7c
(Superior) Planetary subsystem: Design of Fig. 9.4b
Moon phase display device: Designs of Figs. 10.6, 10.8, and 10.10

Model 42

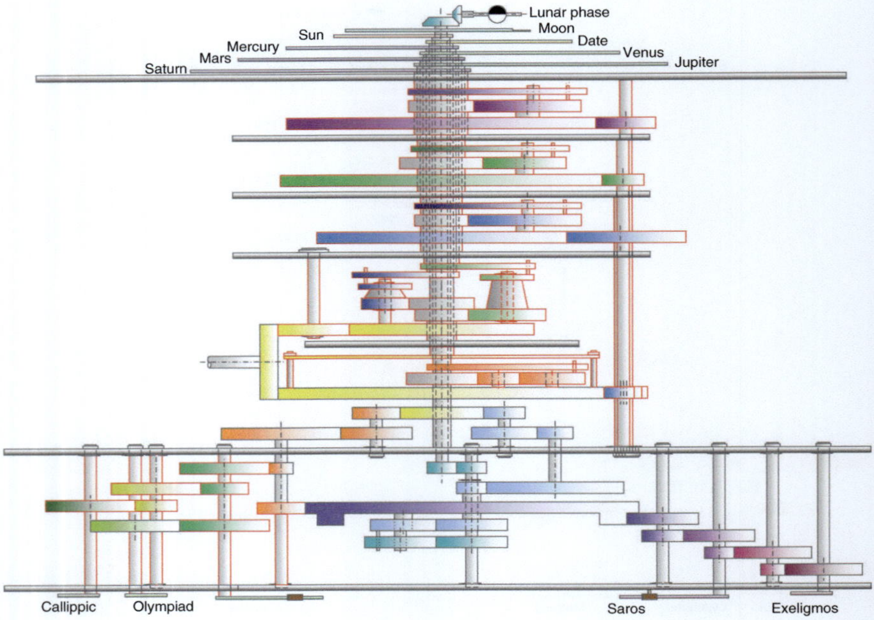

Date subsystem: Design of Fig. 5.7
Eclipse prediction subsystem: Design of Fig. 5.8
Calendrical subsystem: Design of Fig. 6.9b
Lunar subsystem: Design of Fig. 7.9a
Solar subsystem: Design of Fig. 8.14a
(Inferior) Planetary subsystem: Design of Fig. 9.7c
(Superior) Planetary subsystem: Design of Fig. 9.4b
Moon phase display device: Designs of Figs. 10.6, 10.8, and 10.10

Model 43

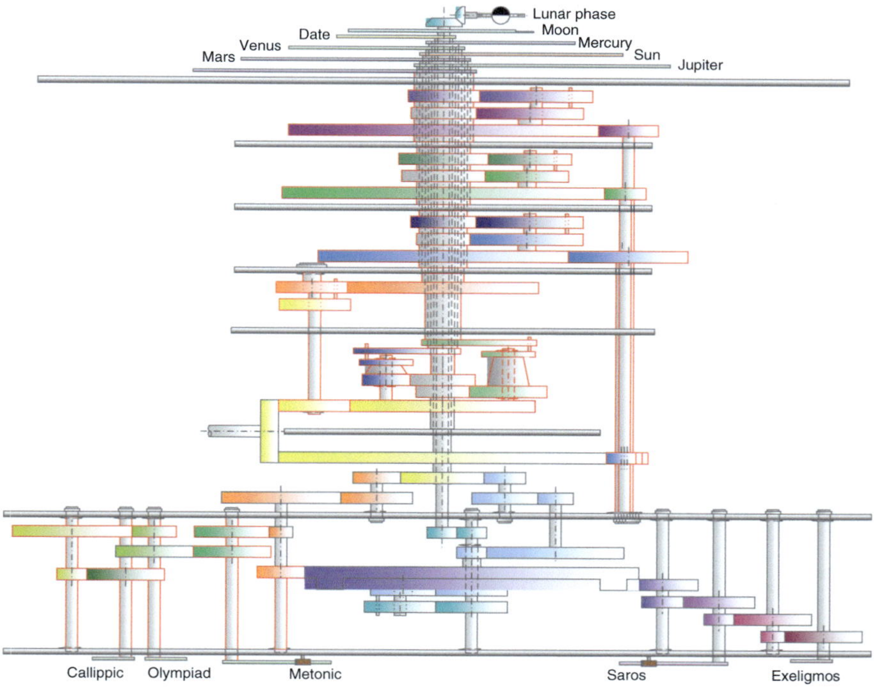

Date subsystem: Design of Fig. 5.7
Eclipse prediction subsystem: Design of Fig. 5.8
Calendrical subsystem: Design of Fig. 6.9a
Lunar subsystem: Design of Fig. 7.9a
Solar subsystem: Design of Fig. 8.5b
(Inferior) Planetary subsystem: Design of Fig. 9.7c
(Superior) Planetary subsystem: Design of Fig. 9.7a
Moon phase display device: Designs of Figs. 10.6, 10.8, and 10.10

Model 44

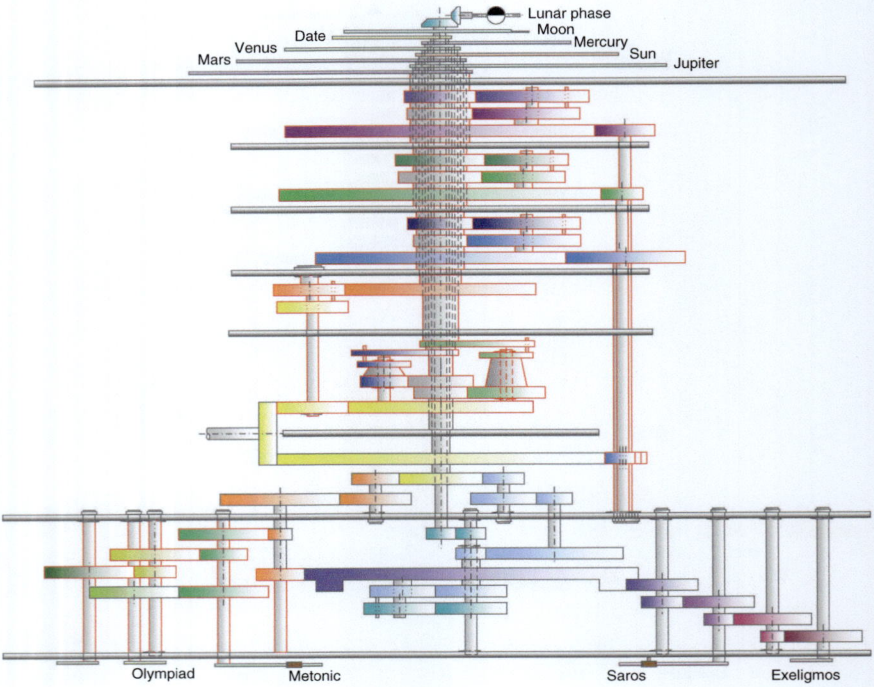

Date subsystem: Design of Fig. 5.7
Eclipse prediction subsystem: Design of Fig. 5.8
Calendrical subsystem: Design of Fig. 6.9b
Lunar subsystem: Design of Fig. 7.9a
Solar subsystem: Design of Fig. 8.5b
(Inferior) Planetary subsystem: Design of Fig. 9.7c
(Superior) Planetary subsystem: Design of Fig. 9.7a
Moon phase display device: Designs of Figs. 10.6, 10.8, and 10.10

Model 45

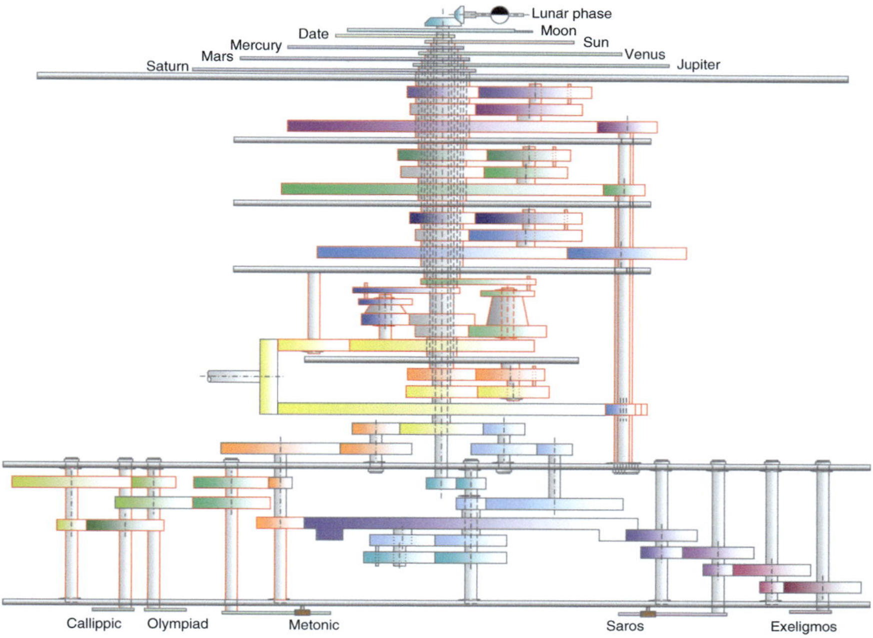

Date subsystem: Design of Fig. 5.7
Eclipse prediction subsystem: Design of Fig. 5.8
Calendrical subsystem: Design of Fig. 6.9a
Lunar subsystem: Design of Fig. 7.9a
Solar subsystem: Design of Fig. 8.9
(Inferior) Planetary subsystem: Design of Fig. 9.7c
(Superior) Planetary subsystem: Design of Fig. 9.7a
Moon phase display device: Designs of Figs. 10.6, 10.8, and 10.10

Model 46

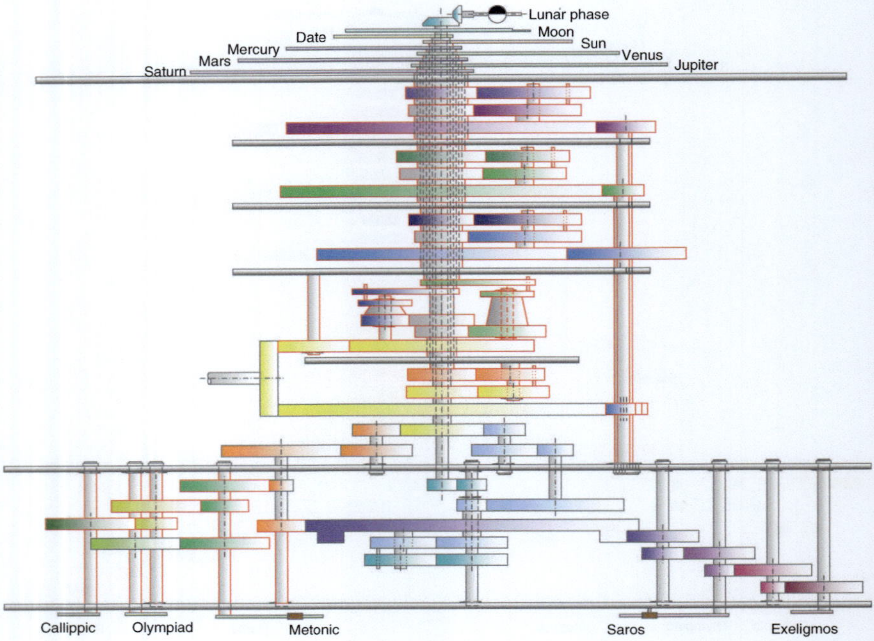

Date subsystem: Design of Fig. 5.7
Eclipse prediction subsystem: Design of Fig. 5.8
Calendrical subsystem: Design of Fig. 6.9b
Lunar subsystem: Design of Fig. 7.9a
Solar subsystem: Design of Fig. 8.9
(Inferior) Planetary subsystem: Design of Fig. 9.7c
(Superior) Planetary subsystem: Design of Fig. 9.7a
Moon phase display device: Designs of Figs. 10.6, 10.8, and 10.10

Model 47

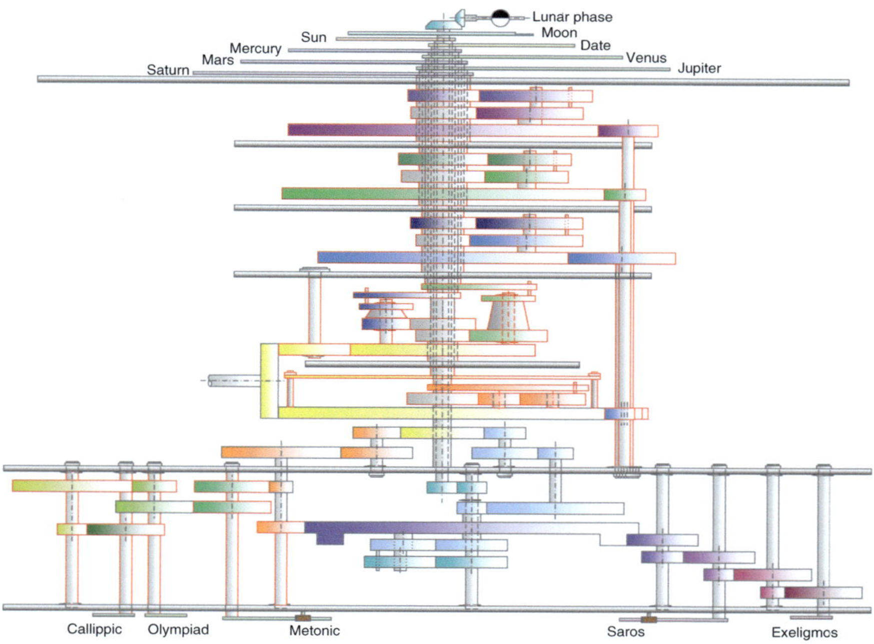

Date subsystem: Design of Fig. 5.7
Eclipse prediction subsystem: Design of Fig. 5.8
Calendrical subsystem: Design of Fig. 6.9a
Lunar subsystem: Design of Fig. 7.9a
Solar subsystem: Design of Fig. 8.14a
(Inferior) Planetary subsystem: Design of Fig. 9.7c
(Superior) Planetary subsystem: Design of Fig. 9.7a
Moon phase display device: Designs of Figs. 10.6, 10.8, and 10.10

Model 48

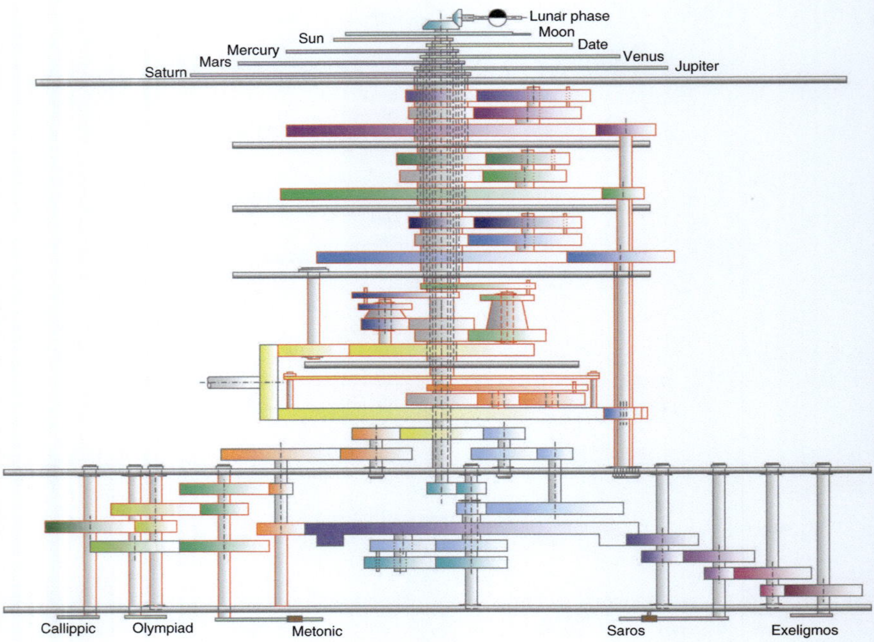

Date subsystem: Design of Fig. 5.7
Eclipse prediction subsystem: Design of Fig. 5.8
Calendrical subsystem: Design of Fig. 6.9b
Lunar subsystem: Design of Fig. 7.9a
Solar subsystem: Design of Fig. 8.14a
(Inferior) Planetary subsystem: Design of Fig. 9.7c
(Superior) Planetary subsystem: Design of Fig. 9.7a
Moon phase display device: Designs of Figs. 10.6, 10.8, and 10.10

Appendix B
Detailed Design of Model 9

No.	Tooth number	Max. radius of gears (center – tooth tip) (mm)	Min. radius of gears (center – tooth root) (mm)	Pin distance (pin position – center of gear) (mm)
1	55	15.7	14.5	
2	100	28.1	26.9	
3	50	15.0	13.8	
4	94	27.6	26.4	11.31
5	215	64.7	63.5	
6	30	9.5	8.3	
7	62	17.7	16.5	
8	60	17.1	15.9	
9	60	17.9	16.6	
10	63	18.7	17.5	9.46
11	215	64.4	63.2	
12	31	9.8	8.6	
13	54	16.1	14.9	
14	94	27.6	26.4	
15	36	10.5	9.3	
16	96	27.0	25.8	9.49
17	215	62.2	61.0	
18	30	9.2	8.0	
19	60	17.9	16.7	
20	60	17.9	16.7	
21	61	18.1	16.9	
22	61	18.1	16.9	
23	Slot	None	None	
24	Slot	None	None	
25	22	6.9	5.7	12.29

(continued)

J.-L. Lin and H.-S. Yan, *Decoding the Mechanisms of Antikythera Astronomical Device*, DOI 10.1007/978-3-662-48447-0

No.	Tooth number	Max. radius of gears (center – tooth tip) (mm)	Min. radius of gears (center – tooth root) (mm)	Pin distance (pin position – center of gear) (mm)
26	40	12.1	10.9	21.5
27	20	6.4	5.2	
28	30	9.2	8.0	
29	30	9.2	8.0	
30	92	27.1	25.9	
31	64	19.0	17.8	
32	224	65.0	63.8	
33	64	15.7	14.9	
34	38	9.0	8.3	
35	53	13.3	12.5	
36	96	24.7	23.6	
37	15	4.5	3.7	
38	27	6.9	5.7	
39	53	14.9	13.7	
40	57	14.9	13.7	
41	60	15.6	14.4	
42	15	4.4	3.2	
43	60	12.8	12.2	
44	57	14.9	13.7	
45	12	3.6	2.4	
46	32	9.3	8.2	
47	32	9.7	8.6	
48	32	7.8	7.1	
49	223	52.4	51.5	
50	188	49.9	49.1	
51	50	13.1	12.2	
52	50	13.3	12.6	7.6
53	50	13.9	12.9	
54	50	14.0	13.1	
55	38	10.3	9.4	
56	48	11.5	10.5	
57	24	6.3	5.1	
58	127	31.7	30.6	
59	53	14.6	13.6	
60	30	8.2	7.4	
61	54	14.4	13.4	
62	20	5.4	4.1	
63	60	14.0	13.0	
64	15	4.2	3.0	
65	60	13.7	12.6	
66	48	14.4	13.2	

Index

J.-L. Lin and H.-S. Yan, *Decoding the Mechanisms of Antikythera Astronomical Device*, DOI 10.1007/978-3-662-48447-0

MIX
Papier aus verantwortungsvollen Quellen
Paper from responsible sources
FSC® C105338

If you have any concerns about our products,
you can contact us on
ProductSafety@springernature.com

In case Publisher is established outside the EU,
the EU authorized representative is:
Springer Nature Customer Service Center GmbH
Europaplatz 3, 69115 Heidelberg, Germany

Printed by Libri Plureos GmbH
in Hamburg, Germany